Growing food in difficult environments

To my son David, waiting for him to have a better world.

To Juan Janna, because of his constant support.

Acknowledgements

I want to thank all those friends, farmers, scientists, environmentalists, professors, indigenous, fans and people interested on new ways to grow food and make it available to everyone.

Vertical Agriculture

Growing food in difficult environments

Vertical Agriculture

Growing food in restricted environments.

By

Orlando Charria

2016

Copyright

Latin Tech Inc

Words from the Author.

All of us would like a better world with peace and equilibrium. Unfortunately, because of the decisions we are making, the planet is facing an increasing danger, which in time, will menace the human population as species.

Among several problems, food supply is one of the most important focuses in current policies of nations, because some research, studies, reports and proofs are showing that if we continue the same path the population is threaten by a starvation disaster.

Have you ever considered the possibility of not having food on your table, even if you can afford to buy or pay for it?

Considering the fact that most of my job and experience has been based on technology, this approach looks for a new way of conceiving agriculture, accepting that not necessarily the path is technological and that there is a lot to do with traditional knowledge. Both concepts the ethno a techno, are explored here, allowing any reader to move from one system to the other.

It is clear that, since the current and future population requires food, the Vertical Agriculture efforts must be directed toward this need.

This book is an invitation to think different, to open your mind and get awareness to stop the way you have been assuming your role in this world, organizing your efforts to guarantee food for future generations.

Contents

Vertical Agriculture

Growing food in difficult environments

Chapter 1

The World we live in.

"But we do have a choice. We can create a prosperous future that provides food, water and energy for the 9 or perhaps 10 billion people who will be sharing the planet in 2050."

Jim P. Leape
Director General, WWF International

This is certainly a beautiful world. From outer space our planet looks like a big blue ball because 70% of the Earth's surface is covered in water. We live in the remaining 30% which is above the sea level.

Our planet has an atmosphere that covers and protects the entire the planet from harmful radiation.

The water is one of our vital resources. Most of the water in the planet is salty (97%) and it's in the oceans. The remaining 3 % is approximately 54994 Km3 of fresh water, found mostly in the polar icecaps and glaciers (69%), ground water (30%) and just a small percentage of 1% can be accessible.

From this small 1% only 0.3% is liquid water available on the surface of the planet: Lakes represent the vast majority (87%), swamps (11%) and rivers (2%)

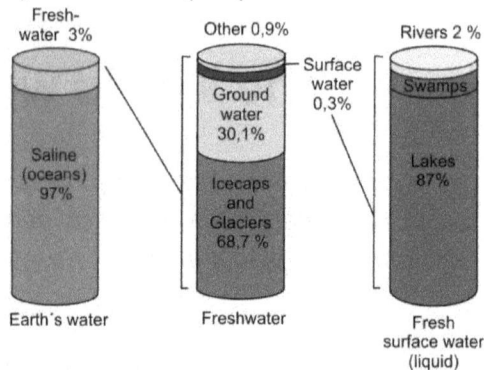

FIG1-1. Source UNEP Vital Water 2010

The icebergs have been melting and breaking. In October 1999, one of 3Km2 separated from the polar cap, affecting the sea level.

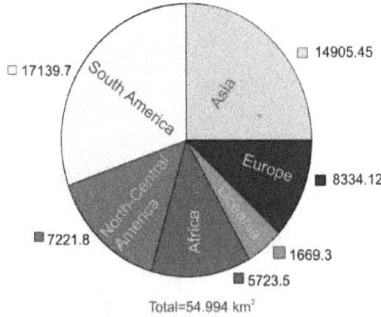

Fig 1-2. Fresh Water in km3. Source Gleick 2004

South America and Asia have the biggest portions of fresh water.

The usage of water can be better understood in the graphic below.

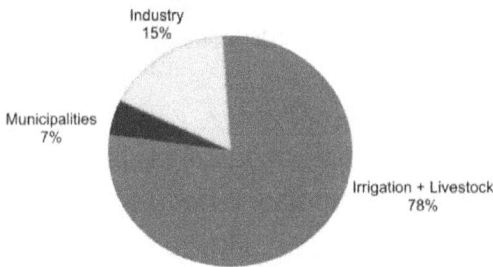

Fig 1-3. Source USGS 2015

Most of the water is used for irrigation and livestock (70%).

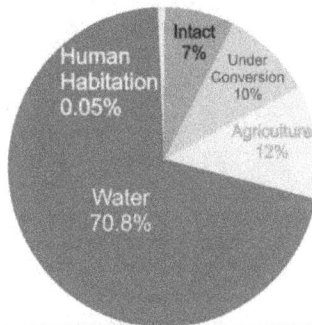

Fig 1-4 Source USGS 2015

Vertical Agriculture

It´s important to know, how water is used inside the homes: Most of the water is used in flushing the toilette, another very important part is used to wash clothes and in the daily shower.

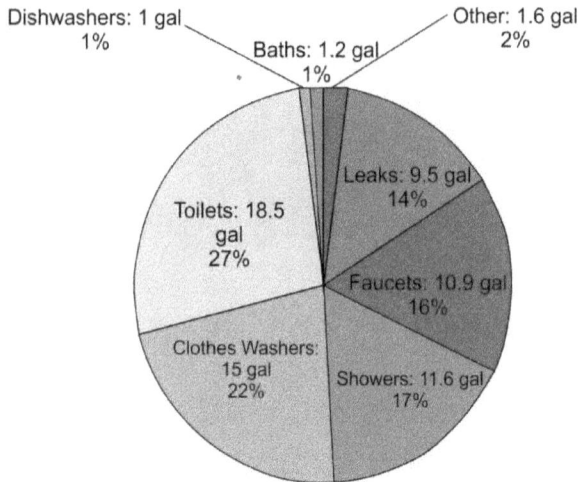

Fig 1-5. Source **FAO**. 2004. Support to the drafting of a national Water Resources Master Plan.

The second most valuable resource we have is earth, where we take most of the elements we need to survive.

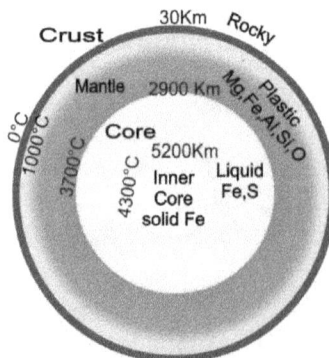

Fig 6. Source pubs.usgs.gov 2015

Growing food in restricted environments

Vertical Agriculture

Earth is mostly formed by 30% of iron which can be found at the core of the planet, 15 % of oxygen mostly found in the earth crust, 15% of silicon and 14% of magnesium.

The usable land on the earth´s surface is used for different purposes, but only 11 % is used to grow food.

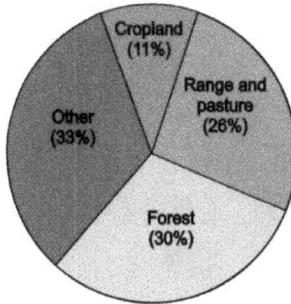

Fig 1-7 World Land Use Source: Wikimedia Commons 2015

Most of the forests (41%) are in the Americas

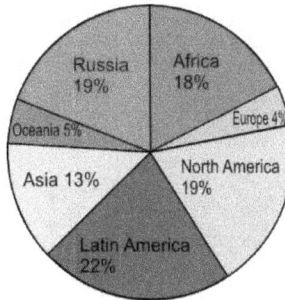

Fig 1-8 World forests. Source: Wikimedia Commons

Every year over one million earthquakes shake the Earth.

The world population has been increasing dramatically in the last years.

In 2015, the number is 7.316 million habitants (usually it's round to 7 billion). In future years the population is expected grow up to levels that can be problematic.

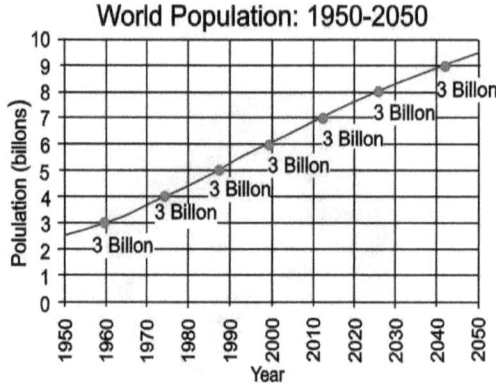

Fig 1-9 Source: US Census Bureau:
International Data Base. June 2011 Update

The continents with higher population growth rate are Asia and Africa.

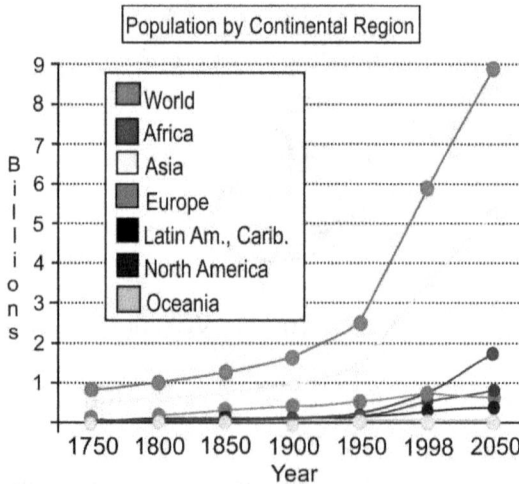

Fig 1-10 Source: UN Population Division: World Population prospects, 1996 Revision

Growing food in restricted environments

Latin America and Europe will grow about at the same rate if nothing changes.

In Asia, China and India are growing their population very fast, compared to the rest of the countries.

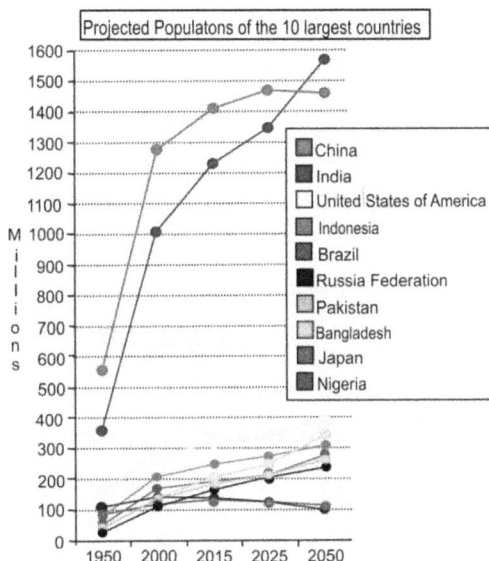

FIG 1-11 Source: UN,Population Division, World Population Prospects, 2000 Revision.

Two billion people live in poverty, more than the population of the entire planet less than 100 years ago. Some population is suffering in misery and starvation in the world in levels higher than ancient times.

In developing countries even with campaigns to limit the number, the couples have more than 3 children even if the income is not enough to support them.

Another vital resource is the energy we get from the sun. It takes 8 minutes 17 seconds for light to travel from the Sun's surface to the Earth. From all the energy we get from the sun, only 51% is reached and absorbed by land and oceans. The rest is mostly reflected by the clouds, the atmosphere and the earth's surface.

Resources:

http://www.gleick.com

Gleick, P.H. 2003. "Global Freshwater Resources: Soft-Path Solutions for the 21st Century." Science, Vol. 302, 28 November, pp. 1524-1528.
http://nebula.wsimg.com/4391a0d8fe2f80e21a8e3b810bd25cc2?AccessKeyId=F22D37E4255246854A5F&disposition=0&alloworigin=1

Gleick, P.H. 1989c. "The implications of global climatic changes for international security." Climatic Change, Vol. 15, No. 1/2, pp. 309-325.
http://nebula.wsimg.com/dc0ca332fd1c090ade283330df39f928?AccessKeyId=F22D37E4255246854A5F&disposition=0&alloworigin=1

World Population Prospects, 2000 Revision
http://www.un.org/esa/population/publications/wpp2000/highlights.pdf

US Census Bureau: International Data Base. June 2011 Update
https://www.census.gov/population/international/data/idb/rel_notes.php?rel=2011-06

US Geological survey
http://www.usgs.gov/

https://fracfocus.org/
estimated_use_of_water_in_the_united_states_in_2005.pdf

Chapter 2

The mistakes we are making.

"We must think not as individuals but as a species"- Professor Brand.

From the movie Insterestellar . Christopher Nolan

Humans are not the only species on the planet, but they are the ones with more effects on others and the environment. Population has been increasing faster than many vital resources, even faster than the renewable ones which use modern technology to speed up the replenishment process.

The future belongs to next generations and it´s unfair to leave them a world with problems that can be solved in current times.
The following are some typical mistakes:

2.1. Waiting for others to make the solutions.
Most of the population is indifferent to the problems and only a minority becomes activist. People don´t like to get involved in so called dead causes, mostly because of the big effort they represent.
After oil spills, the areas are affected in such a way by the oxygen reduction that they turn into a "dead zone" since there are no more support for life. Not much can be said for helping the environment when government decisions are ineffective.

2.2. Thinking that the problems can´t be solved from our perspective or background.
Most of people imagine that to help on the world sustainability and other important environmental issues they must have a related profession like ecologist, biologist, etc. This is a common mistake. Anyone can help, mostly because the work on different environmental problems requires form interdisciplinary efforts.

2.3. Using vital resources as it they were unlimited.
People have no restrictions in consuming resources as long as they have and find food to buy. But what happens if

producers make wrong use of any resource until it gets depleted? How do people get affected by the problem?

As an example, fracking is a practice which uses a lot of water to help on the oil extraction. The resultant polluted water affects the environment and the animals where this extraction method takes place. Several governments have regulations and restriction for the fracking, some others, with less knowledge or perhaps less aware of the consequences.

The rain forests are destroyed without allowing them the time to regenerate.

2.4. Living the present without worrying about the future.

It´s very difficult to predict the future but one thing can be said: Current decisions will have effects in the future. As the butterfly effect, a simple act now can be a bigger problem somewhere else.

As an example, we are over-fishing our oceans, radically changing the species balance in many places.

2.5. Worrying for you and not for the others.

Encouraging nationalism is one of the ways mass people can be controlled and manipulated. This action helps any government in justifying other type of action against others, without any need for approval. When you think as someone born in any nation your thinking is limited by the country you live in. But when you think as race you start considering that you are just part of the world so, all your decisions will affect others.

As humans we are an invasive species since we daily bring to extinction or damage over 50 plants or animals.

2.6. Allowing politicians and commercial forces decide for our future.

Once people voted for a candidate to any government position, they forget the right they have over the elected ones. Politicians should consult and respond to their electors for all the decisions they make, especially for those decisions which can affect people's living. On the other hand, commercial forces use all the power on lobbying to help in changing all type of regulation which can affect their business.

Landfills are the typical form of waste disposal and they are a good example of the way politicians make decisions affecting their community. In the past open dumps were allowed because of the lack of regulations from local authorities. Nowadays, since landfills are part of the environment, they must meet stringent requirements related to the way they are constructed, the leachate management and the way the form part of the society as the main treatment and disposal waste management system. All of these important subjects are parts of the decision making process form politicians.

2.7. Being indifferent to new climate realities.

This is not a trending topic but certainly it should be. People complain about how much cold or hot is the weather but do nothing regarding the problem. The indifference is paid by with more problems. The climate is changing while our local governments and we want to be in denial.

2.8. Being consumers without scruples.

Buying products without any idea on how they were obtained or manufactured is very common among the population. People buy fur coats which come from animals at the border of extinction, diamonds which are full of bloody wars, cellphones, computers and tablets manufactured high contaminating products.

2.9. Living a misinformed life.
We prefer to believe rather than have doubts. We prefer to get the story from someone else rather than doing our own analysis. We get the right information but we have no time to read it, understand, analyze or criticize it.

Fresh water consumption in some countries is 10 times faster than it can be replenished. Human action on soil salinization and erosion several is executed at higher rates than the time that nature could take for restoration.

In order to produce a cow burger 660 gallons of water are needed. This is the average amount of water required to get of the crop form corn, wheat and any other food for the cows. So, in general, producing just one pound of meat will require between 2500 and 5000 gallons of water which is equivalent to the gallons of water a person uses for showering during four months.

Another example of the problem of living misinformed is buying without knowing about the product origin. In the case of pesticides, some products have shown to have more trace of pesticides than others so; special cleaning and care must be taken into consideration when consuming them.

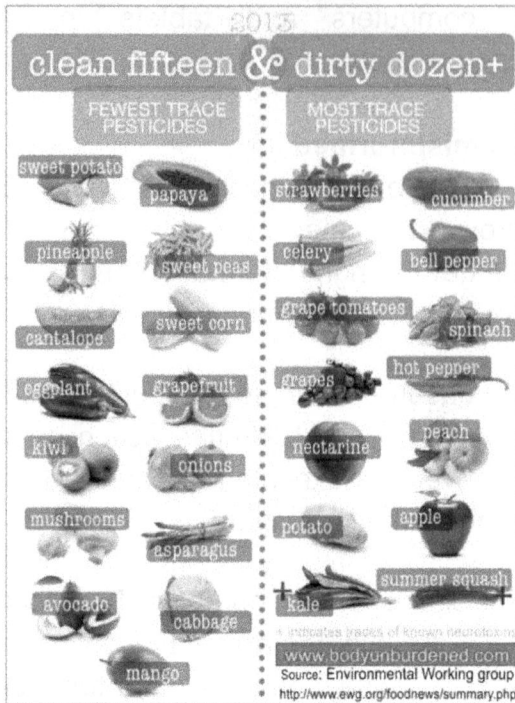

FIG 2-1 Vegetables and Pesticides. Source Environment Working Group.2014

2.10. Considering that technology is the solution.

A big number of the problems can be solved by technology, but there are several cases in which the solution has little or no technology at all. Technology, in some cases, makes the solution unaffordable like it can happen with agriculture where most of the offered alternatives require a relatively high upfront investment in technological equipment, making it difficult for small and beginning farmers

2.11. Generating solid waste without considering its effects.

The Environmental Protection Agency EPA, from USA, has developed a hierarchy ranking for municipal solid waste. This approach seeks to reduce materials use and their associated environmental impacts over their entire life cycles, starting with extraction of natural resources and product design and ending with decisions on recycling or final disposal. The different levels (the greener and higher level the better) look for reducing, reusing and recycling, otherwise known as the three **Rs**.

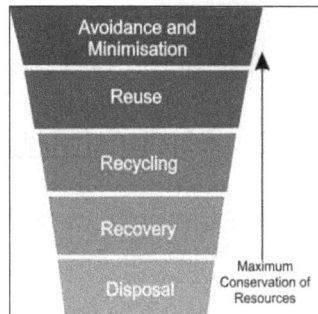

Fig 2.2. Non-Hazardous Waste Management Hierarchy. Source: EPA 2013

Source reduction, also known as waste prevention, or avoidance and minimization, means reducing waste at the source. Aluminum cans apply source reduction because current cans use 1/3 less aluminum than before.

Reuse is the second preferred waste management option and it is the practice of using a material over and over again in its current form but this concept in not applicable in all cases. Disposable products or one-time usage products must be avoided. People are encouraged to use or rent, borrow or buy a secondhand product rather than a new one. For example, use a reusable shopping bag when going to the supermarket instead of taking the ones they give you for free.

Reusing is not true for plastic bottles. A lot of campaigns, most of them from well-intentioned ecologists, wrongly promote the reuse of plastic bottles, for example to grow flowers and other vegetables in gardens, as shown in the next picture.

Fig 2-3. Re-usage of plastic bottles shouldn´t be encouraged.

A simple logic analysis on the subject shows a very terrifying fact:

For every plastic bottle being reused the manufacturer puts another bottle in the market to replace it.

Even with the best engineered solutions, the reuse rate is much smaller than the production rate so reusing plastic bottles is not a solution. The conclusion is that it´s much better to recycle the plastic bottles by forcing the manufacturers to reuse them at the source, on producing part or the total of the new bottles.

Some responsible companies and consumers prefer to use glass bottles which can be returned to the manufacturer to hold products again. A process of cleaning before refilling takes place and then the bottle goes to the market again. The glass bottle can return to the manufacturer several times before it can be discarded because of any damage.Recycling should be the action of taking a product to the place it was manufactured.

The products to be recycled must be collected, sorted and processed to turn some or all the parts into raw materials; these last ones are used to make new or useful products.

Our waste has effects on others, as show on the two photographs below:

Fig 2-4 .Citarum River in Indonesia. Source: Garbage, Waste, Consumption (Landscape and Modernity, Series 7, Benjamin Cohen on June 10, 2009.

Fig 2-5. River of garbage in Manila, Philippines Source: Garbage, Waste, Consumption (Landscape and Modernity, Series 7, Benjamin Cohen) on June 10, 2009.

2.12. Encouraging wrong help models and ineffective initiatives.

Several organizations spend millions of dollars to help the world in problems like poverty, hunger and disasters. Most of these processes have proven to be long, inefficient and susceptible of corruption.

For example, when a town experience water scarcity, some governments encourage population to plant on their gardens vegetable species with less water requirements like cactus or vegetation which is common in desserts, instead of attacking the real problem and the water waste. Is this a real solution?

In some countries, most of the help comes with training the local people to self-sustainable. This is a very typical situation since although it has been found that the model is accepted by communities but in some cases and show certain index of success, time, generations and change in local governments has negative effects on both current and future results.

2.13 Allowing a non-environmentally responsible urbanism.

As population grows cities require more space, building and facilities. The problem is that most civil engineers and architects, who are mostly responsible for planning the cities' growth, don´t take into consideration collateral effects of their decisions. As an example, Miami and California have planted palms instead of trees. This decision, just made by practical reasons, has contributions to raise the temperature. Palms don´t offer the same advantages of trees in producing shadow, water retention and helping with moisture and CO_2 to prevent any cement jungle from absorbing the heat from sunlight. Fruit trees would be preferable and if the roots and falling fruits become a problem, humanity have to learn how to make solutions to each of them.

2.14 Being indifferent to Planned Obsolescence.

Most of people never ask how long a product will last. Several manufacturers know that and take advantage of your innocence, by selling you a product that will fail within certain amount of time (deliberately planned time) for you to buy a new unit. The reasons are technology, fashion or simple damage of the device.

There are several supporters of the idea that shortening the lifespan of a product helps the economy. Some others care about the waste and costs for population considering these actions as illegal.

There is a big difference between your decision and the manufacturer´s decision to force you to buy a new product. There are several cases you have no choice: printers, software, cars, machines, light bulbs, etc.

The case is so critical that you can find a lot of info about commercial cases , the light bulb for instance, that can last as many years as wanted but the manufacturers decided by themselves to shorten the lifespan of the bulb. There is case in Livermore, California, where the lifespan of a lamp has reached 114 years and still counting.

Resources:

Environmental working group
www.ewg.org

Environmental Protection Agency USA
www.epa.gov

Benjamin Cohen
Garbage, Waste, Consumption (Landscape and Modernity, Series 7)
http://scienceblogs.com/worldsfair/2009/06/10/garbage-waste-consumption-land/

Long lasting light bulb
http://www.centennialbulb.org/

Chapter 3

The impending disasters.

"I only feel angry when I see waste. When I see people throwing away things we could use." —*Mother Teresa*

3.1 Starvation

Is it conceivable to have worldwide places without food? Is hunger something we can't eradicate?

According to some reports from World Food Programmed (WFP), one person in eight on the planet goes to bed hungry each night and one child out of three is underweight.

On its efforts to eradicate hunger, the organization has identified, among many reasons, six of very high relevance: Poverty trap, war and displacements, unstable markets, lack of investment on agriculture, climate and weather, and food wastage.

Perhaps the last three subjects are more important according to the focus of this book.

Hunger and undernourishment (also called malnutrition) have similar meaning. According to FAO, Undernourishment is a synonym of hunger and means that a person is not able to get enough food to meet the daily minimum dietary energy requirements, over a period of one year.

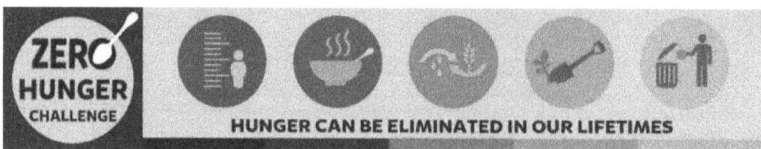

Fig 3-1. ZHC. Promotional strip. UN 2015

The Zero hunger Challenge is a program form United Nations to boost up food security, following the precepts of the Millennium Development Goals (MDG).

Assume that you decide to do something simple. Take seeds of an edible vegetable and plant them wherever there is a

2

green place, garden, road, farm, patio, park, etc. There is a big chance that several of those seeds become trees or plants with fruits. Nature has always been wise and some species develop and grow even without any care. This is why wild vegetables with fruits can be found in places where there is no human intervention at all. Then, why do we need to wait to produce food. Why do we need to pay for food if somehow we could find it easily wherever we go? Why we spend thousands of dollars in ornamental plants when we can do the same with edible plants and they can be also ornamental? Is this the answer for food security?

3.2. Water scarcity

New scenarios of climate change are changing the water distribution all over the world. Some places are experiencing extreme droughts which never happened before. The rain has been changing in frequency and intensity in several countries affecting the way life was normally carried on.

little or no mover scarcity
ltrysical water scarcity
approaching physical water scarcity
economic water scarcity
not estimated

Fig 3-2. Water Scarcity. Source UN World water

Water scarcity is then an impending disaster. Every person must have the right to have access to water, but in reality it doesn't happen and water availability, not necessarily potable, is becoming a problem for all the countries, no matter the economic level.

3

In some countries the amount of water is not the problem, but treatment plants and the access to potable water. Drinking water directly from water sources like rivers, lagoons and seas is a risk with very important effects on human health.
Mining industry, oil extraction with fracking techniques, landfill leachates, oil spills in the ocean are contaminating the waters in a way that simple water treatments are ineffective, leading to uncontrollable diseases.

According to the amounts of water used for agriculture, the production can be affected. Heavy rains have harmful effects on exposed crops, especially when the rain drops become hail and fall on places where farmers don't expect them at all. Droughts have negative effects on crops which in time become an economic and social problem.

California in USA is experiencing the biggest drought in a millennium. Although it is considered the eight biggest economies of the planet, money has nothing to do when it relates to water.

Simple activities like taking showers, washing your hands, cooking, and washing of clothes are problematic in some places, because there is no water for such purpose other than bottled water.

A good example is the unprecedented fact of what happened in 2015 in California, USA. A water scarcity of 98% in the whole state forced the local government to adopt measurements and penalties of 10.000 US dollars to individuals, institutions and companies which were wasting water. The initial goal was to save 25% of total water consumption for the whole state. A call for stopping the irrigation of gardens, car washing, and shower duration, restrictions on toilette flushing and other activities was made.

A very important remark is the fact that at that particular time in California, agriculture and animal husbandry had an 80% of the total water consumption, while the remaining 20% was related just to home and businesses.

3.3 Consumption habits

The consumption habits principally from industrialized countries are generating risks of an imminent disaster for the rest of the world.

There is a very strong link between consumption, poverty, inequality and environment and since this is the only plant we have, it´s a problem which should concern to all.

It is necessary to help low income consumers to grow their economy, to be more conscious on amounts of goods to be consumed and the origin, end and production technologies applied. Every action of consumption has environmental concerns.

According to the research conducted by the Pacific Institute in Oakland, an organization dedicated to environmental studies, the residents of California use more than 1500 gallons of water a day, because of the meat and dairy products consumption. Indirectly they are consuming around 900 gallons of water since this is the amount required to produce every pound of cheese. 1000 gallons of water are needed to produce one gallon of milk.

Consumption is much related to obesity. This is one of the main problems affecting world population since over-eating represents more than normal consumption. It doesn´t matter if the food is healthy or not or if there is a sedentary lifestyle, because obesity represents the need for more food production and availability. Off course, there are some minor exceptions as obesity, in addition to the health problem, can be based on poorer food quality which is available in both developed and developing countries.

From consumption another bigger problem is derived: food waste.

A lot of food is not consumed and therefore it is wasted. A big responsibility belongs to people that discard food which has been already processed and served.

Annual food waste by region (Kg/person)

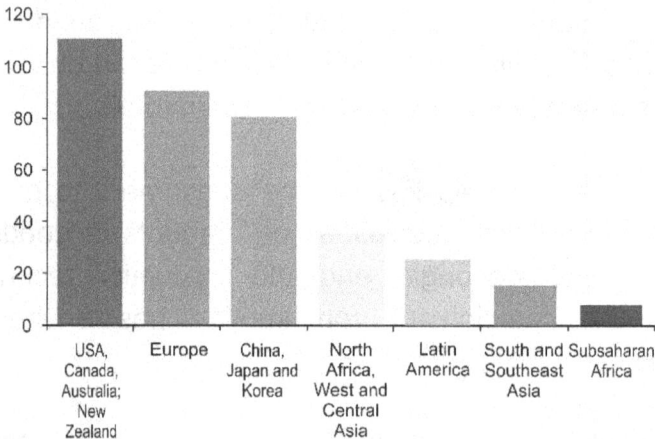

Fig 3.3 Source Gustavson et al.(2011) FAO Via Shrinkthatfootprint.com.

Estimate of per person food waste for seven different regions of the globe, and is based on the 2011 FAO report, Global Food Losses and Food Waste.

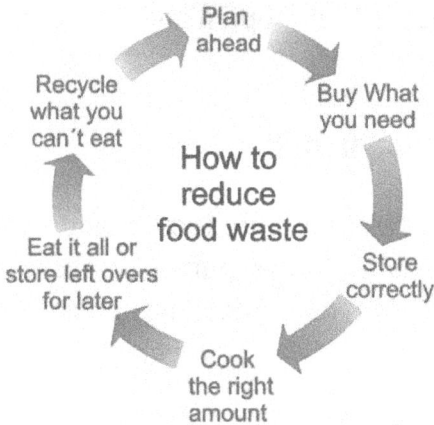

Fig 3-4 Source Somerset Waste Partnership 2015

Fig 3-5 Source, Leonard 2015 Whiskey city

There is a real need for reducing per-capita resource use.

3.4 Food production methods

If we continue with the same rates on current production methods, the new realities, like climate change, will affect the food security chain and this will create a disaster to future generations.

The improvements for food production methods have to be addressed in two ways:

Food production optimization and reduction of production losses.

3.4.1 Food production optimization

Develop more efficient food production and distribution without taking over more land. In this particular case, as you will read later on this book, vertical agriculture, could be one very good alternative.

3.4.2 Reduction of production losses

The food´s supply chain implies a lot of stages between the harvesting process until the food reaches the retailers and the consumers. During all the independent processes a lot of losses are present.

In 2011, FAO estimated the losses of food production and the food waste at consumer side. The facts were impressive, from the graph below you can notice the almost 40% of roots, tubers, fruits and vegetables are not consumed but lost.

Although cereals have less production losses, they are almost double compared to the consumer waste.

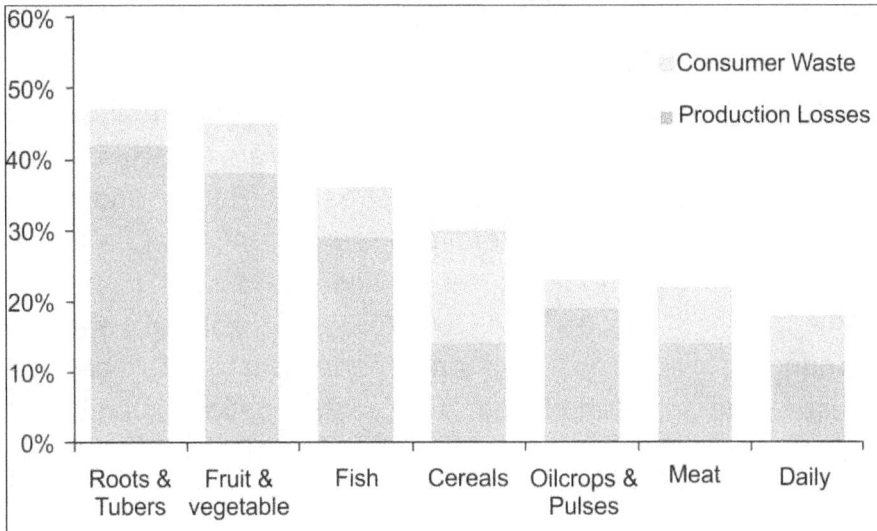

Fig 3-6. Source Gustavson et al.(2011) FAO Via Shrinkthatfootprint.com

In general, every year nearly 940 million tons of food is lost during production, which is about 25% of total world food supply.

Because of the inefficiency in the food supply chain, just few countries have adopted relevant policies to reduce food waste. Recently, the European Parliament decided to work a resolution to achieve in 2020 a 50% food waste reduction, compared to the one in 2012.

Several food retailers and brands of the food supply chain joined the initiative to support the waste reduction for all the stages they were involved in.

3.5 Pollution

Human activities imply pollution. The problem is the uncontrolled manner it is made.

Vertical Agriculture

The oil industry uses much questioned extraction methods: one of them is called "fracking".

In the process of extraction of oil and gas, huge amounts of water, with added sand and chemicals are injected to oil reservoirs using high pressure streams. Rocks and earth beneath the surface are literally "dissolved" to generate a path for the oil and gas to get to the surface.

This method, in addition to the huge polluted water it produces, weakens the lands and the place become susceptible to earthquakes. The contamination of land and water destroy any near ecosystem and that's a high price to pay.

This method uses nearly 1000 billions of gallons of water just in USA and 35 % are used just in California, which is curiously, one of the states with the most extreme droughts.

California is a very interesting case, since for many years this state had no major limitations for the use of underground water in many applications. Big corporations, like the ones in the oil industry and high scale agriculture, used the water affecting most of the small farmers. This unrestricted use of water has led to a huge water scarcity problem which has forced the local government to take actions affecting all the local population and business.

References:

Zero Hunger Challenge.
http://www.un.org/en/zerohunger/#&panel1-1
FAO
http://www.fao.org/docrep/014/mb060e/mb060e.pdf

Global Food Losses and Food Waste.

http://www.fao.org/ag/ags/ags-division/publications/publication/en/?dyna_fef%5buid%5d=74045

Love Food Hate Waste community.
http://www.somersetwaste.gov.uk/more/food/lfhw-community/

Have a taste of waste food.
http://thewhiskeycity.com/2015/05/have-a-taste-of-food-waste/

What is fracking and why is it controversial?
http://www.bbc.com/news/uk-14432401

Chapter 4

Interdisciplinary Research

Knowledge is extracted from a fully integrated world. Knowledge is "disintegrated" by disciplinary units called departments in universities. How can knowledge, discovery and dissemination be re-integrated?
Denise Caruso and Diana Rhoten, Ph.D., Hybrid Vigor Institute.2001

Environment is an integrating concept and study object for social, political, economic and ecological issues. It allows the formulation of concepts which are the basis for the generation of knowledge that follow the holistic and systemic principles from environmental sciences.

Environment is a very interesting focus for any research on agriculture with the specific goal of food (auto) sustainability at any scale.

It is also important the understanding of the real need of how innovation on agriculture should applied and developed.

In general, as in most of the fields of knowledge, in agriculture there are two basic types of research:

- **Applied Research:** The main goal of this type of research is to develop the all the basis needed to support the scientific knowledge in relation to diversity and complexity on all the selected fields of study in Agriculture. The problems arise in the great difficulty in describing and explaining the variability of specific processes.

- **Pure or Basic research:** The objective is to improve the state of art for all the knowledge on agriculture. The big problem is based on the fact that there is certain difficulty in identifying, describing and explaining an object of research trough models and methodologies.

4.1 Interdisciplinary and Transdisciplinary projects.

There are many ways to interact within projects related with agro –ecosystems

	Monodisciplinary	Multidisciplinary	Interdisciplinary	Transdiciplinary
Object	Single and individual	Unique for each discipline	Common	Common
Thinking	Simple	Simple	Complex	Complex
Work	Personal	Personal /Group	Integrated	Integrated
Results	Individual	Individual/Múltiple	Integral	Integral
Autonomy	Total	Total	Systemic	Systemic
Interaction	None	At the end	Continuous	Continuous

Fig 4-1.Table of Parameters for Research. Source Charria 2015

Most of the research on agriculture should be interdisciplinary, because the vision is less demanding than transdisciplinary research.

Academic research tends to be directed toward these last two; however, doctoral research is more focused on interdisciplinary efforts.

Most of researchers always explore different approaches on the kind of research to be made on agriculture in order to clearly define who is going to be responsible in defining the objectives from each research activity. There is certain academic freedom which allows the researcher to define his own topics on his research work. He also tends to resist any intervention from advisors, mentors or colleagues. The subject of the topics is selected in such a way that it must satisfy his curiosity, criteria and relevance. In other cases, they look for scientific reward which helps them to promote his career.

Working on interdisciplinary research requires a background on any discipline and the knowledge, obtained by experience or specific study, on the subject or problem to research.

Some authors have pointed out the main situations when an interdisciplinary research is required:

- When the knowledge about a problem is uncertain.
- When there is no consensus about the nature of the problem.
- When the problem represents a real need that affects population and the ones in charge of solving it.

Some renowned entities like the National Science Foundation (NSF) from USA, have acknowledged the value of interdisciplinary research to promote knowledge on the different fields and to speed up the scientific discoveries and processes, because very often the ideas transcend the goals of a single discipline or program.

The NSF also strive for integrating research and education through interdisciplinary training in order to prepare a work force able to overcome any scientific challenge, using innovation as fundament.

Interdisciplinary research could be defined as the conceptual and practical integration of more than one discipline in order to find solution to complex problems which require the collaborative and participative work of several disciplines and that share a common goal.

Transdisciplinary concepts are a wider approach of interdisciplinary research focused on developing participative actions which must result in a more holistic achievement. There is a transgression of disciplinary boundaries with the purpose of understanding, defining and solving complex problems which integrate and transform emerging knowledge.

4.2 How interdisciplinary research on agriculture should be?

It depends on the particular subject to be studied. In order to try to provide an answer two basic conditions are assumed as required:

- It is recommended to avoid a single perspective. It´s necessary the holistic vision by involving of other researchers from other disciplines.
- The work should be more comprehensive, thru collaborative and not individual work, when having groups. This condition represent a challenge for researching leaders who are responsible for training of group members, direction of meetings and establishing the goals

Several experiences have shown that research on traditional agriculture, produce better results, with more effectiveness and less costs, when the focus is not only the improvement of a single process but the correct combination of improved agricultural practices and all type of efforts driven by interdisciplinary research.

Research on sustainable agriculture or sustainable development is always related to multi or interdisciplinarity. Although agriculture production can be modeled like just a physical process, the true is that every research implies an agriculture zone which is at the same time used for experiments, tests, business and in many cases, as a home. Environmental sciences are not enough to research the field, because other social, technical, economic and political dimensions are present.

Vertical Agriculture

When performing research on agriculture, there are six main key factors to make research activities on agricultural systems: Environmental impact, handling on resources used for agriculture production systems, increasing production and profitability which implies better or innovative technologies for crop production, education and training on sustainability, attitudes and perceptions and last ,but not less, synergistic collaboration.

Agricultural practices have positively evolved throughout the time, possibly because of all the research efforts on fundamental knowledge, proper training and spreading of experiences and results to small farmers. These last ones can grow their own agricultural concepts by summing up technological with traditional knowledge.

One of the ways to understand the proposal of vertical agriculture, on this book, is focusing on assuming this situation as a good opportunity and new way to develop knowledge on agriculture, through interaction with all possible actors and the real appropriation and respect of ALL types of concepts, considering the ones found on traditional and technological visions with the same level of importance.

This is a really good opportunity for interdisciplinary research to develop and enhance environmental knowledge under the new simple directives of Vertical Agriculture in search of beneficial and innovative solutions on food production.

One of the main mistakes which can be corrected, under this new exercise of knowledge development, is recognizing that there are more types of knowledge, other than the ones based on pure sciences and technologies. The advantage of involving different parts of population, like small farmers, and

6

giving them the chance to be heard, will provide a new sense of confidence that the solution will be successfully embraced and applied.

It is worth to mention that not all problems in agriculture require an interdisciplinary approximation, but undoubtedly, when applied, will lead to much better results. Disciplinary research could require a later integration with other disciplines, especially when there are social and economic impacts which go against public interests. For instance, if small farmers are forced to increase production while minimizing waste in agricultural practices could lead to a very expensive solution.

It has been found that one of the aspects to enforce interdisciplinary research is the funding of projects and fellowships with co-authorship work. The express condition of including several disciplines for interdisciplinary research as one of the criteria to define if the project deserves to be funded is a good way to enhance the communication and collaborative work when performing early stages of the planning of the project.

In Germany, an interdisciplinary research on Precision Agriculture (www.preagro.de) looks for increasing the efficiency on agricultural practices while promoting a better environmental compatibility. The central goal is focused on obtaining the best economic yield from the arable land available, respecting all the principles for good agricultural practices with environmental responsibility. The institution is interested in showing the importance of interdisciplinary research among all the involved partners team, which include thirteen research institutions, two service and two software companies. Additionally, sixteen farms and their staff are part of the project

The correct interrelation between technological and not technological actors, between users and developers of technology, between producers and environmental institutions, between governments and population, clearly show the best way of using interdisciplinary research to solve a complex environmental problem.

It is quite known that agriculture guarantees subsistance through production of food and raw materials. Traditionally, habitants of any country depend on agriculture production. Modernization and commerce (national and international) had reduced the dependecy on local cultivated or home products, but new scenarios of climate are becoming problematic, even for import of food, in cases of low or no availabilty of local crops.

Vertical agriculture, as a new approach which can be carried out in an interdisciplinary manner, could help in developing alternative solutions for low cost food production.

Environment can be considered as an integrating agent and a field of study for social, economic and ecological issues. This condition enables the Environmental Sciences for creation of knowledge under holistic and systemic principles, for any system type on Agriculture, mainly because of the need for sustainability on any scale.

References:

Hybrid Vigor Institute
http://hybridvigor.org/

NSF
Introduction to Interdisciplinary Research
https://www.nsf.gov/od/oia/additional_resources/interdisciplina
ry_research/

Precision Agriculture
www.preagro.de

Chapter 5

Agriculture

"It is time to rethink how we grow, share and consume our food. If done right, agriculture, forestry and fisheries can provide nutritious food for all and generate decent incomes, while supporting people-centered rural development and protecting the environment. But right now, our soils, freshwater, oceans, forests and biodiversity are being rapidly degraded. Climate change is putting even more pressure on the resources we depend on, increasing risks associated with disasters such as droughts and floods. Many rural women and men can no longer make ends meet on their land, forcing them to migrate to cities in search of opportunities. A profound change of the global food and agriculture system is needed if we are to nourish today's 925 million hungry and the additional 2 billion people expected by 2050."

Resolution adopted by the General Assembly on 27 July 2012 [without reference to a Main Committee (A/66/L.56)] 66/288. The future we want.
http://www.un.org/ga/search/view_doc.asp?symbol=A/RES/66/288&Lang=E

Agriculture is one of the main activities carried out by the population through all the history. It has played a key role in the way the world has developed and sustained so far.

The world agriculture is an adaptation to English language of the Latin term "Agricultura", which is formed by two different words: "Ager" which means "Field or Land" and "Cultura" which means "Cultivation".

The cultivation of land is defined in English also like the "act, process or art of tillage of the land or soil"

In English language, Agriculture has synonyms like Farming and Husbandry so, several publications and authors prefer to use the term Farming. The problem arises when trying to make a translation into any other language since the word Farming practically doesn't exist and has to be translated as Agriculture. So, the reflection is to think about of using more universal terms and because of this reasoning, this book is named "Vertical Agriculture"

The word "Farming" is a noun originated from the verb and noun "Farm", which in the sense of agriculture began around 1719 and it is related to an area to cultivate the land and raise animals. The conversion from verb to noun or adjective is only common in English but not always it is possible in most of the rest of the languages of the world.

In 1798, without knowing and considering the current technological advances, the prediction of the famous demographer Thomas Malthus, had stated that while food production tends to increase arithmetically, population tends to increase naturally at a (faster) geometric rate, which could lead the world to a starvation disaster.

Growing food in restricted environments

Vertical Agriculture

Agriculture is the production of food and other goods, raise of domesticated animals (livestock), and cultivation, maintaining and developing of forests (also known as forestry). The study of agriculture is known as agricultural science.

For many years Agriculture was the main work for people, allowing the development of what it´s called the traditional agriculture. This kind of agricultural practice had a lot of productivity problems until a new era of new methods appeared ad spread among the population.

One of them, the method for synthesizing ammonium nitrate changed the way cultivation was made because it increased enormously all the yields for agriculture. It made that the labor for recycling nutrients by rotating crops, the longtime processes and animal manure became less necessary.

Between 1940 and 1970, a process for improving agricultural results called the Green Revolution took place. Norman Borlaug is considered the "Father of Green Revolution" because he was the one who leaded the team in charge of formulating the methods, technologies and policies that helped the world to produce more food and so, preventing any future starvation.

The Green revolution´s work was targeted to increase global yields on agriculture by the scientific development of high yielding varieties of common staple grains such us corn, rice, wheat and other cereals. The work also included several other fields like correct and better irrigation techniques, training on use of pesticides and synthetic fertilizers like the nitrogen and very important, increasing the availability among the farmers of hybridized seeds which could assure a better production.

Growing food in restricted environments

Vertical Agriculture

A very important remark is that the knowledge was not only applied in developed countries but transferred to most of the developing world.

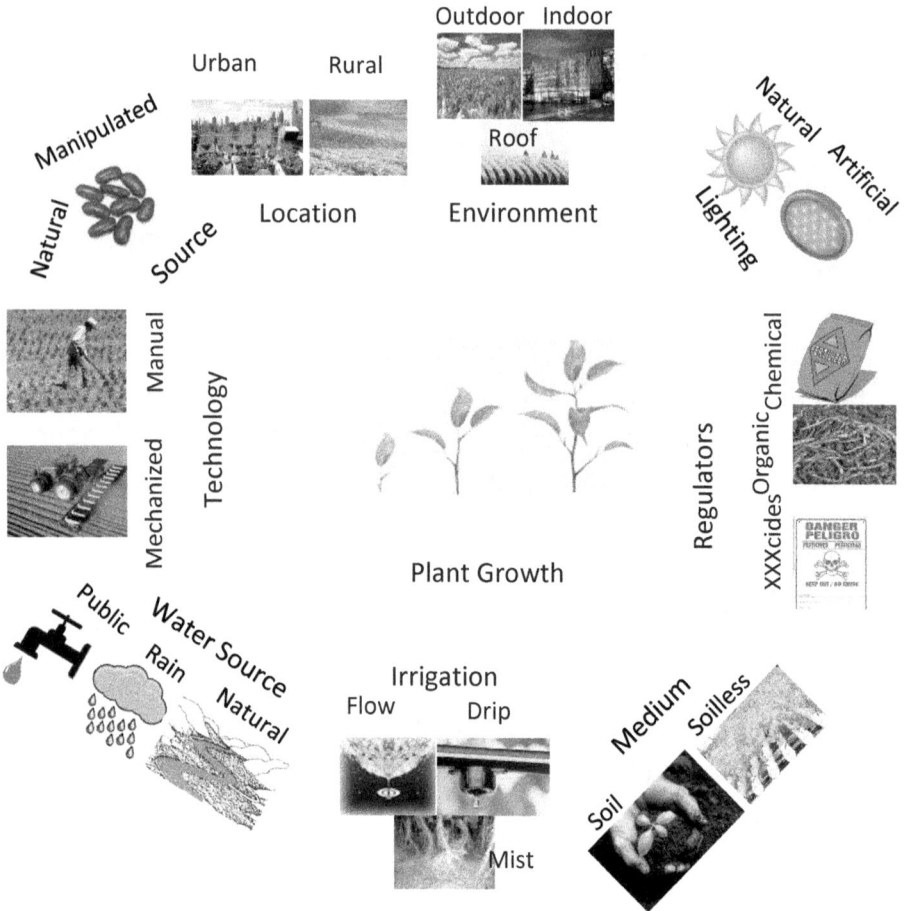

Outdoor Indoor
Urban Rural
Manipulated
Natural
Roof
Source
Location
Environment
Natural Artificial
Lighting
Manual
Mechanized
Technology
Plant Growth
Regulators
XXXcides Organic Chemical
Public Water Source
Rain
Natural
Irrigation
Flow Drip
Medium Soilless
Soil
Mist

Fig 5-1. Agriculture In The World. Charria 2015

Although the world has an agriculture which can be considered as of much better productivity, many technologies available, modern irrigation techniques, highly efficient mechanization and extraction equipment, there is a lot of

Growing food in restricted environments

research and work to do. Even in these days Agriculture is always facing and generating problems.

Farmers experience high vulnerability to climate events like droughts and rains. Several processes, including agriculture are generating waste, pollution and water contamination.

To succeed the Millennium Goals we need to solve problems on climate change; lack and degradation of arable land, water scarcity and other factors have made a call of urgency in developing new agriculture methodologies.

Conventional agricultural methods to grow food outdoors have proven to be problematic. With a growing population competing for resources like land and water, the current agricultural practices are proven to be if no problematic, at least inefficient.

The new agriculture must produce food with minimum resources of water, nutrients and space, less labor and intervention, minimum exposure to weather, no contact with soil and less usage of pesticides, herbicides and other agrichemicals.

Food production should be less problematic, with more chances to have successful crops, even if they are small, with more accurate and predicted results. New agricultural practices must allow an easier way to know, predict and account any possible success or failure of food production.

It is important to mention that although most of the authors consider agriculture as a blessing which helped the development of civilizations, some other authors have stated opposite ideas. The popular Discover Magazine published an

Growing food in restricted environments

article from the researcher Jared Diamond (May 1987, pp. 64-66) named *"The Worst Mistake in the History of the Human Race"*.

In his own words, the author has written the following regarding agriculture as the catastrophe from which the civilization never recovered:

"To science we owe dramatic changes in our smug self-image. Astronomy taught us that our earth isn't the center of the universe but merely one of billions of heavenly bodies. From biology we learned that we weren't specially created by God but evolved along with millions of other species. Now archeology is demolishing another sacred belief: that human history over the past million years has been a long tale of progress. In particular, recent discoveries suggest that the adoption of agriculture, supposedly our most decisive step toward a better life, was in many ways a catastrophe from which we have never recovered. With agriculture came the gross social and sexual inequality, the disease and despotism that curse our existence."

Since the agriculture was developed when people stopped being hunters and gatherers and became farmers, the criticism to think if this was a good decision, is easily supported. In this century, there are habitants like Bushmen or nomads that still live from hunting and gathering. Compared to farmers, they have more advantages because the work less hours a day, have no major living problems, have more leisure time, have more balanced and nutritive diets and seem to be reluctant to adopt agriculture or become farmers.

While for hunters and gatherers starvation never was or has been a reality, through history this has been the case for many farmers with agricultural knowledge. This situation provides a

Growing food in restricted environments

good reflection to consider if future agriculture could take advantage of this experience and generate methods for people limit the labor on food production

New methods should pay attention to both ways of agricultural knowledge, no matter if it is developed in either urban or rural centers. Some knowledge refers to the ancient ways of agriculture, developed and based on traditions, with very intensive human labor, related mainly to small farmers.

Some other knowledge uses technology and automation to control and optimize food production: it can include materials, nutrients, irrigation, aero and hydroponics and mechanized extraction methods.

Technology can be useful on agriculture, but there is still a lot of research and work which can be done without it, by giving and recognizing the importance to the ancient knowledge.

New production systems are required to achieve more predictable results of food production, which is also known as accountability in agriculture, possibly they have to be different than the outdoor crops and techniques and its associated vulnerability.

An appendix regarding different types of agriculture has been added to the book in order to keep the purpose of this chapter.

Conventional agriculture is too vulnerable to climate change because the crops are exposed to external weathers. Since the new extreme climate events are becoming more frequent it is a need to understand the importance of developing agricultural practices in indoor environments especially in countries where seasons are present. In other cases, droughts and heavy rains represent the main menace to crops so less demanding protections could be required and simple roofs or

Growing food in restricted environments

covers could be not enough but a simpler and lower cost solution.

Indoor Agriculture, is the same practice of cultivation of crops and raising of animals, similar to the conventional agriculture, but with the difference that all the activities are carried out inside a containing room or clsed environement such as greenhouses, warehouses, rooms, climate chambers and phytotrons.

Restricting the environment can have both advantages and disadvantages.

These closed and artificially controlled environments can be located anywhere. If located in urban centers, food like vegetables can be produced fresh and with very low transportation costs, because they are close to the consumers. The production can be off season because of the weather independency. Some other added values could be found in the possibility of having less usage of harmful chemicals like pesticides, herbicides and fungicides. From the ponit of view of vital resources like water, it can be said that indoor systems can be more efficient in using them.

But growing food indoors can have some up-front high costs which have been decreasing because of the new, promising and scale production of the technological equipment involved. These environments require control systems for heat/cooling, lighting, irrigation and power, in addition to the place preparation and adaptation costs. The costs are also affected by the type of product and its availability, but some indoor farmers have found ways and models to make it a profitable business.

Growing food in restricted environments

There is no major interest to compete with outdoor agriculture or conventional agriculture, but the idea of generating a market able to keep a sustainable consumption with more ecological and healthy standards. Recent climate events are evidencing the need for a new agriculture, not necessariky indoors but protected, and the possibility of limiting the risks on food production and availabilty.

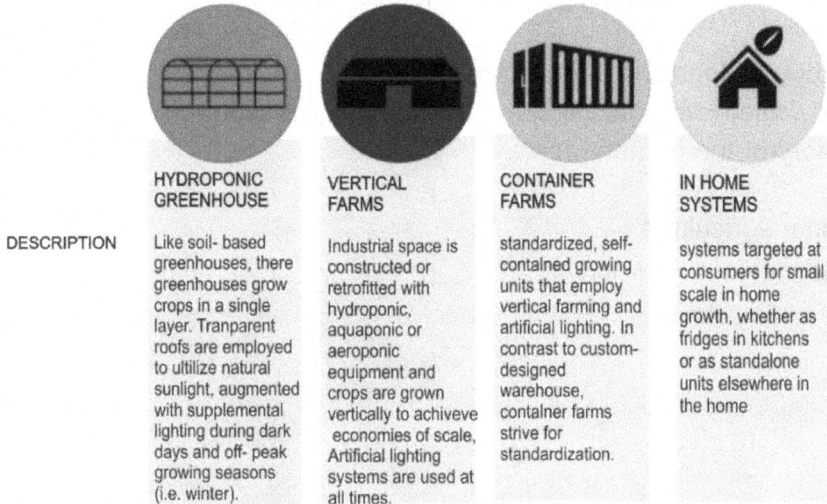

	HYDROPONIC GREENHOUSE	VERTICAL FARMS	CONTAINER FARMS	IN HOME SYSTEMS
DESCRIPTION	Like soil- based greenhouses, there greenhouses grow crops in a single layer. Tranparent roofs are employed to ultilize natural sunlight, augmented with supplemental lighting during dark days and off- peak growing seasons (i.e. winter).	Industrial space is constructed or retrofitted with hydroponic, aquaponic or aeroponic equipment and crops are grown vertically to achieve economies of scale, Artificial lighting systems are used at all times.	standardized, self-contained growing units that employ vertical farming and artificial lighting. In contrast to custom-designed warehouse, container farms strive for standardization.	systems targeted at consumers for small scale in home growth, whether as fridges in kitchens or as standalone units elsewhere in the home

Fig 5-2- Source Indoor Agriculture 2015

Growing food in restricted environments

References:

UN. The future we want.
http://www.un.org/ga/search/view_doc.asp?symbol=A/RES/66/288&
Lang=E

National Geographic. Green Revolution
http://www.nationalgeographic.com/foodfeatures/green-revolution/

Discover Magazine. Agriculture
http://discovermagazine.com/1987/may/02-the-worst-mistake-in-the-history-of-the-human-race

Indoor agriculture
https://indoor.ag/whitepaper/

Chapter 6

Growing food indoors

"It is a concept whose premise is easy to envision: Stack up "hi-tech" greenhouses on top of each other and locate these "super" indoor Farms inside the urban landscape, close to where most of us have chosen to live. However, I came to realize early on that making happen will not be an easily attainable goal, and certainly not simple from an engineering and design perspective."

From the book "The Vertical Farm" Dr Dickson Despommier 2010.

As mentioned before, the term "farming", which comes from the noun "farm", doesn´t exist in several languages so it is translated as agriculture. This is the reason some authors prefer to use agriculture.

It is important to mention that the word farm is referred as the place where several agricultural activities take place so the expression "Vertical Farm" can have a more precise translation into most of the languages.

6.1 Vertical Farming. The early concept.

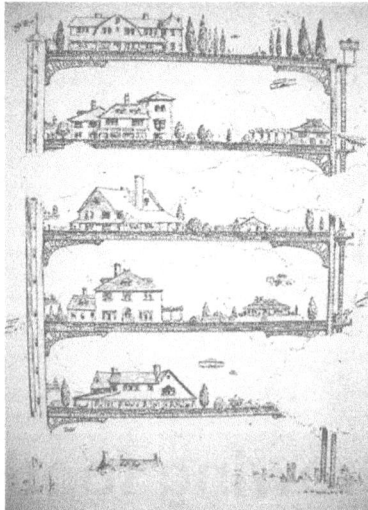

Fig 6-1. Life Magazine 1909. Delirious Rem Koolhaas

The original idea of was coined in 1915 by Gillbert Ellis Bailey in his book "Vertical Farming". His main idea was to increase the available space for farming by going down as much as possible using explosives.
In his own words referring to the farmer and the possibility of more cultivating land:

Growing food in restricted environments

" What is wanted is something to enable him to move vertically down and double his acreage, and double his yield, by doubling the fertility of the soil, by doubling the depth of the feeding zone, by doubling the water supply, by cultivating the ground to double and treble, the former depths. " Vertical Farming" to coin a name, is the keynote of a new agriculture that has come to stay, for inexpensive explosives to enable the farmer to farm deeper, to go down to increase his acreage and to secure larger crops. Instead of spreading out over more land he concentrates in less land and becomes an intensive rather than an extensive agriculturist, and soon learns that it is more profitable to double the depth of his fertile land than double the area of his holdings, and he learns that his best aid and servant in this work is a good explosive."
(Segment extracted from Pages 67and 68)

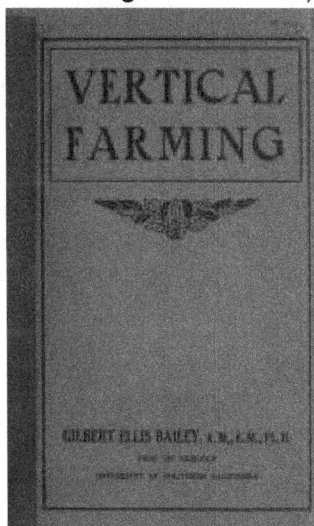

Although the idea was going down, rather than up, it is pretty clear that the concept is to make better use of land for food production. For many years traditional agriculture continued its development in the many ways we know today, but none of the reasons were related to scarcity of vital resources.

John Hix in his book "The Glass House" has written a complete survey regarding the development of glass enclosures, from the mid 16Th century up to now, as the way to create artificial micro-climates, which allow the possibility of growing any type of plants in hostile environments.

The new greenhouses are becoming very popular, because they are affordable, scalable and include a lot of features which are needed for indoor production: fertilization and irrigation systems, controls and equipment for lighting, heating and cooling. Some manufacturers offer turnkey solutions and custom designs.

6.2. Vertical Farming. The modern concept

The innovative buildings are not conceived for a commercial purpose but as way for the community who lives there to develop some of their own food.

In 1999, PhD Dickson Despommier, professor of Microbiology and Public Health at Columbia University, found very powerful reasons of ecology and food sustainability to suggest the implementation of what he calls "Vertical Farms", as a solution for food production affected by several environmental problems.

He considers that it is quite possible to develop urban farms, by stacking high technology greenhouses one on top of the other like floors in buildings. By having food production within the city will solve several problems to prevent the world form a future starvation disaster.

In theory, controlled environments should help to produced vegetables with fewer risks, toxicity and with lower effects of carbon footprint.

NASA has shown high interest in research for self-contained systems capable of producing food, which must work with limited resources of water and space, to make then suitable for long duration space travels.

Some skeptical people don´t believe that this approach can be competitive. The high costs involved in these technological greenhouses could lead to a higher price for this "premium" production.

In his book, The Vertical Farm: Feeding the World in the 21st Century, readers are taken on a journey inside the vertical farm, where multistory buildings are filled with fruits and vegetables grown locally for entire cities.

6.3. A vertical Farm Project

The Deustches Center for Lufft and Raumfhart DLR, from the German Aerospace Center, using a Controlled Environmental Agriculture Technologies, has developed a 37 floor Vertical farm with interdisciplinary engineering design.
The importance of this project relies in the fact that it´s not an adaptation of any building but a complete design which started with specific designs from the beginning.

It is a 31 floor building for food production and 5 floors underground for water waste management and fish farming. Different vegetable are grown in the floors but floors 8, 18 and 30 which are reserved for environmental control.

Growing food in restricted environments

All the final processing, selling and delivery are done at first floor and the basement.

This project is a good approach to grow food indoors in a vertical way

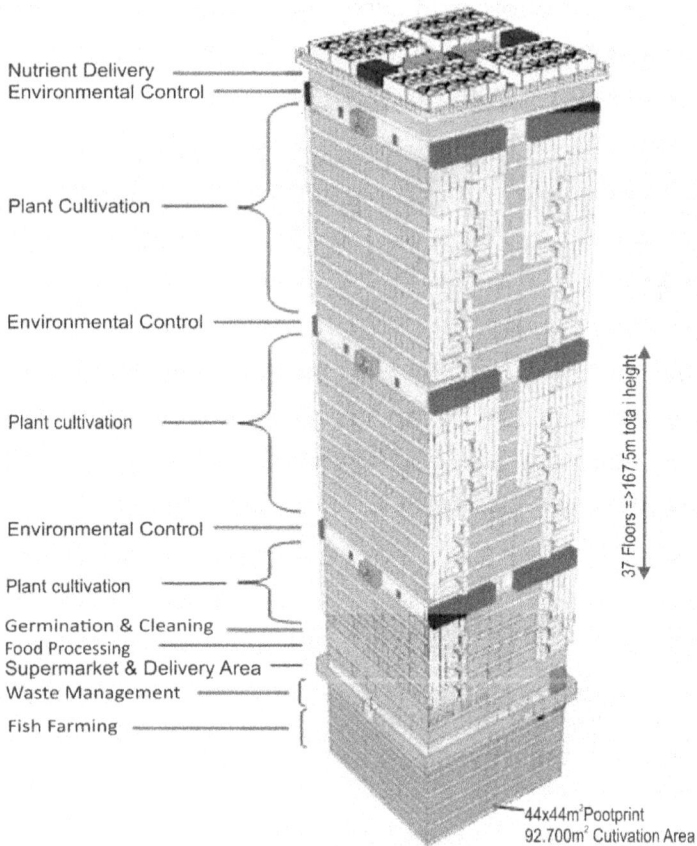

Nutrient Delivery
Environmental Control

Plant Cultivation

Environmental Control

Plant cultivation

Environmental Control

Plant cultivation
Germination & Cleaning
Food Processing
Supermarket & Delivery Area
Waste Management

Fish Farming

37 Floors =>167,5m tota i height

44x44m² Pootprint
92.700m² Cutivation Area

Fig 6-2 Source Poster ASC. Schubert Vertical Farm*. 2014

In the lower underground levels, the fish can be feed with some of the sub-products, to complement the nutritional load of any community.

Growing food in restricted environments

Vertical Agriculture

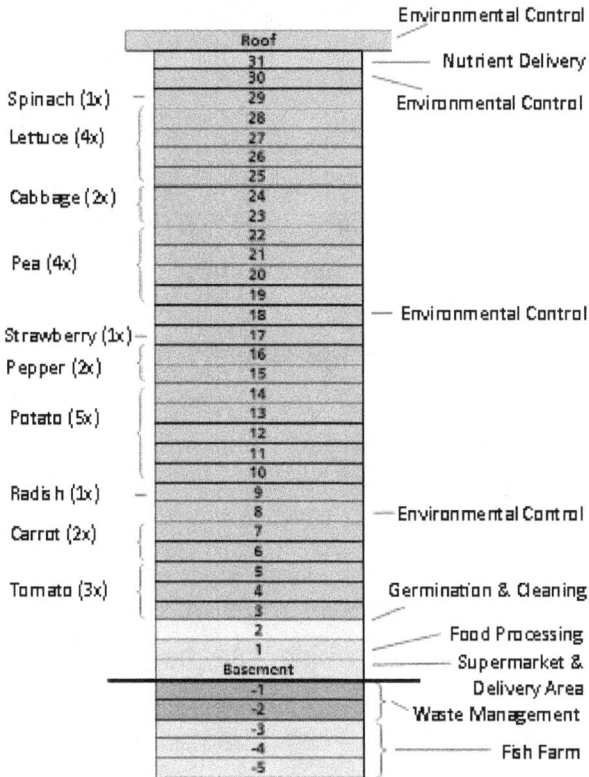

Fig 6-3. Source Poster ASC. Schubert Vertical Farm*. 2014

The building, with a 44x44 m2 of footprint, has a growth area of 9,27Ha, which is a smaller extension if compared to the equivalent 215,87Ha, require in conventional (horizontal) agriculture.

6.4 Vertically Integrated Greenhouses

Another way of having food production in urban centers is based on using agriculture for innovative food production on wall facades and roofs of buildings as part of the bioclimatic design.

Growing food in restricted environments

The concept has been defined and applied since 2007 by ,
Kiss + Cathcart with Arup, Engineers and Bright Farm
Systems as a:

"...highly productive, lightweight, modular, climatically
responsive vegetable culture system designed to be installed
in the curtain wall of a high rise building".

The Vertically Integrated Greenhouse (VIG) is a technology for
building and constructions which combines several concepts
like special double skin façade construction, hydroponics and
food production.

The Double Skin Facade (DSF) is a very innovative strategy in
architecture for the construction of buildings with
environmental concepts, which can help to save energy in
nearly 30%. It is based on the fact that the envelope of a
building, roofs and walls, get affected by external climate
conditions.
Adding systems, like ventilation windows, shading devices and
vegetable production for thermal insulation, which can
respond to ambient conditions, will bring several benefits like
the need for less cooling inside the building, better comfort
because of natural light throughout the year and a very good
acoustic insulation.

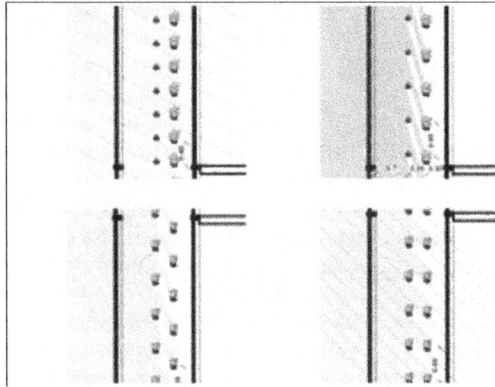

Fig 6-4.VIG sections showing daylight control. Source
Kiss + Cathcart 2014

Since there is a cavity between the two walls, the heat accumulation is used in different ways during summer or winter seasons.

In summer the heat flows from the lower part of the building and goes outside through automated windows. In winter these windows remain closed and then the heat recirculates in the cavity. Both cases help with the comfort and the savings of heating/cooling energy of the whole building.

Fig 6-5. A Vertically Integrated Greenhouse in a double skin office façade. Source Kiss+Cathcart

Growing food in restricted environments

Vertical Agriculture

The combination of plants, light and building designs offer a very interesting ways of saving energy since heating and cooling is made by using the sunlight throughout the year within the DSF, which in turn brigs comfort to the occupants.

Taking advantage of the solar radiation and sun light, the design operates along with the HVAC systems in the building to decrease the electric energy bill, by using electric air conditioning only when it´s really needed.

The food production is made on hydroponic trays which are hanging on a cable system. Complete tray rows can be inclined to maximize solar absorption and provide shade and insulation to the interior. During summer, since windows can remain open, the fresh air can circulate freely providing very good ventilation for the place. In winter the windows are closed and the heat on the plants, by solar absorption is sent out thru some ducts on top.

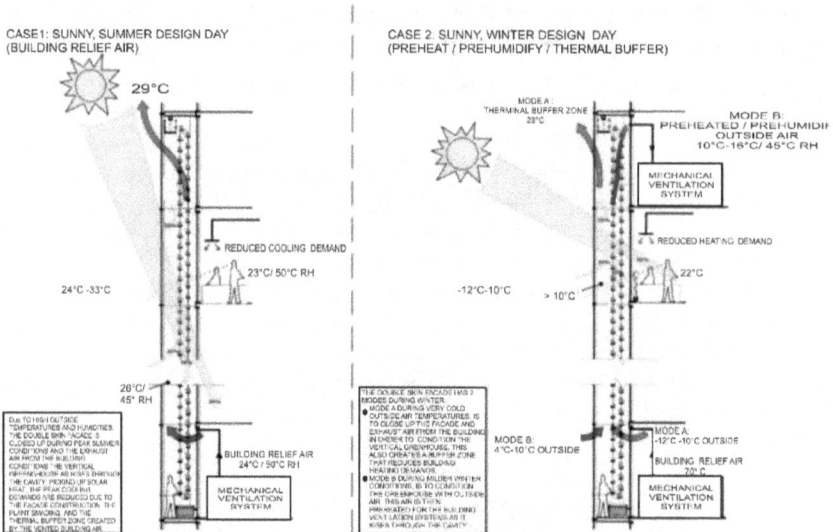

Fig 6-6 Integration of the VIG with building systems.
Source Kiss+Cathcart

Growing food in restricted environments

The costs of new technologies like this one can be high and despite of the obvious advantages, the implementation of the systems of VIG has been limited. It is only a matter of time for this approach to become popular.

6.5 Vertical Hydroponic Systems.

Valcent, a company with locations in Texas, Vancouver and Cornwall, has developed a vertically stacked hydroponic tray as a very innovative food production system which guarantees the maximum absorption of sunlight since the plant trays move on rails.

One of the applications is the cultivation of lettuce and some other vegetables to feed the animals in Paington Zoo in Devon.

Valcent has a commercial system called Verticrop (see picture below)

Fig 6-7 Verticrop tray system.

Growing food in restricted environments

Vertical Agriculture

Source Verticrop (www.verticrop.com)

The designers point out several differences or advantages of their system could provide when compared to field soil agriculture:

- The energy needed is very little. One year crop of 500.000 lettuces would require just the same energy as a desktop computer running 10 hour a day. Traditional agriculture would require seven times more energy to have the same yields.
- Since the trays are moving, the plants receive the same amount of natural light.
- The required are is much smaller than the field agriculture, in a factor of 20.
- The water required could be up to 8% of the total water used for soil agriculture.

The main idea of bringing these systems to produce food in urban centers has positive effects on health, carbon footprint, energy and other environmental costs.

Technologies in hydroponics, like the one Valcent has developed working along with vertically integrated greenhouses, could help in solving the sustainability problem of food production, growing population, nutrition and food safety.

Growing food in restricted environments

Vertical Agriculture

Fig 6-8. Plant Project. Source: Piotr Szpryngwald's Plant System.

One of the main problems of space travels is precisely growing food in restricted enviroments and limited resources. Mirko Ihrig, a german industrial designer from Lund University in Sweeden , winner of the Core award in 2012 for the project PLANT. This project was developed along with NASA Space Center in Houston, to grow fresh food in space as a complement of astronauts´diet and to research on plant growth for zero or very low gravities. The appeareance, taste and freshness of these vegetables is much better than the dried-freezed bars astronauts are used to eat.

The small re-usable pillows are filled with a grainy content which helps to support the roots, supply the nutrients and keeps the moisture. The lower wall of the pillow is a permeable membrane that allows the pass of water sucking it from a bed to the interior of the pillow.

A simple tool opens a small hole on top to accommodate the seed of the plant.

This experience will help to plan the future on food supply for long duration space travels, moon bases, space stations or distant outer space explorations.

References:

Bailey, Gillbert Ellis, Vertical Farming
http://www.biodiversitylibrary.org/item/71044#page/3/mode/1up

Growing food in restricted environments

Vertical Agriculture

Despommier_Dickson
The Vertical Farm: Feeding the World in the 21st Century
ISBN 978-031261139-2

Indoor Agriculture
http://agfundernews.com/3-big-challenges-for-indoor-agriculture4864.html/

Schubert Vertical Farm
http://www.agrospaceconference.com/wp-content/uploads/2014/06/Poster_ASC_2014_Schubert-Vertical-Farm.pdf

Vertically Integrated Greenhouse
http://www.kisscathcart.com/integrated_agriculture.html

Verticrop
www.verticrop.com

PLANT
http://www.szpryngwald.com/

Growing food in restricted environments

Chapter 7

Vertical Agriculture

"We stress the importance of access by all countries to environmentally sound technologies, new knowledge, know-how and expertise. We further stress the importance of cooperative action on technology innovation, research and development." United Nations Resolution adopted by the General Assembly on 27 July 2012 . 288. The future we want.

http://www.un.org/ga/search/view_doc.asp?symbol=A/RES/66/288&Lang=E

Imagine that we have the chance to develop a new agriculture, based on what we know so far and the fact that we are aware about the problems our practice will produce for humanity and the land. In addition, we already know the risks the agriculture is facing.

Acting responsibly, we have to develop certain rules, take advantage of the interdisciplinary knowledge and try to spread this new vision as much as we can. Vertical Agriculture is a call to act different, fighting against paradigms, to re-think about the ways we grow our vegetable species, either for consumption or for commerce.

Vertical Agriculture is a new proposal, to produce food, using the land in an intensive way (vertical) rather the extensive (horizontal), using limited or scarce vital resources like land or water.

It is important to clarify that this approach has many common concepts and also big differences with the original ideas from George Ellis Bailys and PhD Dickson Despommier.

7.1 Proposal for Vertical Agriculture

Conventional agriculture can be developed in rural and urban places with or without sophisticated technological tools. Industrialized countries can afford the use of technology to ensure food production, but reality shows that most of the world use traditional agriculture and very simple tools for the agricultural practice.

Before starting any analysis, we have to re-define several concepts, which will help in understanding this approach. The first important remark arises when the orientation of the crop is taken into consideration. If the crop is on just one level and with extensive usage of land, the agriculture can be considered as Horizontal. Some other considerations can be added if the crop is located outdoors or indoors, buts this assumption is not relevant for now.

On the other hand, if the agricultural processes are based on very simple rules (defined and justified later), but they try to be implemented on one or more levels above the land using all resources as restricted, under this approach this practice can be considered as a Vertical Agriculture.

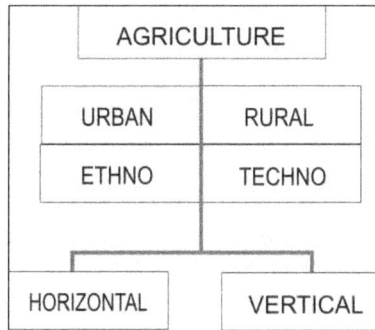

Fig 7-1. Proposed Conception for Agriculture

The graphic above, shows that the agriculture can be practiced in urban or rural places, with or without technology and has two possibles paths: Horizontal, the classical way to see it,and Vertical, as a new proposal, which works with the same principles but under some basic and new rules.

It is important to mention that despite other approaches, the use of technology is only part of the solution, because this vision of vertical agriculture considers that there are many other low cost alternatives can be explored.

Growing food in restricted environments

Parameter	Horizontal Agriculture	Vertical Agriculture
Climate Change	Vulnerable	Prepared
Crop distribution	Extensive	Intensive
Land and Water	Uses them as resources	Uses them wisely
Knowledge	Ancestral and technological	To be developed. *
Crop	Vegetables	Edibles

Fig 7-2. Comparison chart of basic parameters.

*Some other visions imply the use of technology

This proposal looks for a methodology to develop a cultural transformation toward Vertical Agriculture to achieve less vulnerability to climate and more sustainable use of land, nutrients and water, with easier feedback and more predictable production results.

7.2 The core of this vision

The main intention is to develop a new approach for agriculture, where traditional (horizontal) agriculture practices can have a smooth transition to a different and probably more efficient way (vertical).

Since this is also a learning process, it won´t be an imposed but a guided, participative and shared process where the community will have an active role to guarantee the social appropriation and the development of new knowledge.

The final objective is to provide a cultural transformation which can turn the traditional practitioners of agriculture (for example a horizontal farmer), with all their vulnerability and experience into a new and into more prepared farmer (vertical farmer) which has understood the needs for an adaptation of agriculture to climate change, develops and share knowledge with others and has better control of his livelihood and nutrition.

Just imagine if in a new future you could ask:

Are you a horizontal or a vertical farmer?

7.3 Are we prepared for VA?

The initial answer is yes. We have the knowledge and, if needed, the technology for most of the technical aspects. Of course it´s not an easy task but helping the population to work on the subject will speed up the process. Several barriers have to be turned down:
- The governments and their indifference to the current climate situation.
- The cultural challenge to adopt new practices.
- The massification of knowledge, technologies and products.
- The usage of resources and the need to define limits.
- The inertia to accomplish new projects and goals
- The need for economical support to extend this approach.

7.4 Is VA for everyone?
Food is a need for everyone. The scale of food production is not considered important as long as population can have food on the table. When prices of food go beyond a threshold where limitations for buying become a generalized situation, people and governments will realize of the importance of developing ways of agriculture at every scale or for everyone.
Growing food in restricted environments

If food is scarce, it doesn't matter if people have money to buy, big extensions of land to plant or even the best technology to make food production.

It is quite simple to understand, a big scale climate event could lead the world to a starvation disaster of unpredictable results.

Currently, there are more people interested in growing their own vegetables for different reasons: small home gardens, community production, hobby, fresh food, etc. Size and technology for a vertical agriculture practice is not the problem. Different types of Vertical growers can co-exist, from simply fans which grow one plant at home to industrial producers which have a business.

One of the main problems of the fans is that new homes or apartments don't have much room to cultivate and in the future this situation will become worst. For urban habitants, they don't have time, space, the technology and perhaps nor the interest to grow food for business but for themselves.

Farmers, mostly located on rural lands, have the space and use a given technology to grow several types of vegetables. They also face different risks: pricing of agrichemicals, vulnerability to environmental conditions and the cost of required technologies. This group could act as a partner of large scale growers but hardly compete with them.

The last group is formed by large scale growers and farmers with applied technology. Most of them only produce one type of vegetable. Since the crops are based on industrial processes, the need to keep the pace, very often is in the boundary of environmental responsibility and the danger to

Growing food in restricted environments

ecosystems. In this group you can count oil palm, corn, sugar cane growers, etc.

7.5 How classify all the different vertical growers.

In order to categorize any vertical agriculture practice, it´s necessary to establish two possible paths: Location and Knowledge. It is necessary to know where the agriculture takes place and the kind of required knowledge to be applied.

7.5.1 According to Location

Similar to the traditional concepts of agriculture, the vertical agriculture practice can be located in urban or rural places. When its located in the boundaries of the cities it could also be called peri-urban agriculture. Vertical Urban agriculture can be developed in a very small apartment located in a given downtown up to a big building within the urban center which uses some floors for food production.

Nowadays it's very common to find cultivation in urban and peri-urban locations.

7.5.2 According to the applied knowledge.

This is one of the main differences with traditional agriculture and other proposals, because Vertical Agriculture must be considered a new approach or trend and not a new technology. Vertical agriculture represents a good opportunity to develop and get knowledge at every level.

The intention is that, opposite to what happens with traditional or indoor agriculture, a vertical grower can develop the same or more knowledge than high technological producers.

Two types of knowledge are considered here:

Ethno-knowledge and Techno-knowledge.

The first one, Ethno-Knowledge is proposed for the traditional farmers, families or persons with simple and not very technological cultivation. The vertical ethno growers, similar to the traditional agriculture growers, conceive vegetable production with the use of soil and manual (or low scale automated) irrigation. This is the respect and recognition for those persons, in many countries, who have contributed to the development of agriculture in all the ways we know today. A knowledge that is worth to rescue, embrace and promote.

The Techno-Knowledge, on the other hand, allows and propends for the use of technology in all possible ways but following the simple criteria defined for Vertical agriculture. The advantage of having control on every possible risk and unattended crop production could lead to very promising, sustainable yet profitable systems.

For Vertical techno-knowledge growers, soil is not necessarily required since other ways of plant fixation can used or developed in the worst case. High technological irrigation techniques for water and nutrients, like the available ones in precision agriculture, can be used, reducing human intervention at its minimum. Complex sensing and control systems for agricultural production can be required and of course allowed and encouraged.

One of the main goals is that units and systems used to grow food on vertical agriculture must look about the same for ethno and techno versions.

7.6 Vertical Agriculture Growth Units (VAGUs)

The Vertical agriculture must be understood and conceived, as a way to produce food but with certainly simple but clear concepts, defined as the "commandments" or rules of gold, explained later.

Not every vertical disposition of vegetables should be considered as Vertical Agriculture. In the same sense, not every vertical garden or production should be catalogued within this description as long it fulfills some basic requirements.

A vertical growth system is required in order to apply, understand and promote the concepts in many ways, simplifying the efforts and costs. The name for these initial vertical crop structures is VAGU (from Vertical Agriculture Growth Units).

A VAGU represents a unitary and countable cultivation system and also it can be associated with a concept of "mini-land" or "portable land". These basic units are low costs assemblies which can be hung on or attached to any internal home wall and that can grow different vegetables without much care and without any effect of climate conditions.
They can be irrigated manually or through an automatic control box which takes care of the solution with nutrients to feed the roots

As for now, according to the technology available on them, two types of VAGU can be implemented:
Ethno and Techno

Similar to the concepts exposed before for Ethno and Techno knowledge, the VAGUs Ethno use soil and have very simple technology since water irrigation is manually made. In contrast, VAGUs Techno can include all types of technologies

Growing food in restricted environments

available in soil-less agriculture, automation, natural or artificial light and water control and irrigation.

Some very basic types can be implemented but one important requirement is that the VAGUs Ethno and Techno should preferably share the same size and appearance. Every time, there is a new VAGU let´s say Ethno, it´s necessary to conceive the Techno version and viceversa
Both types, for now, look about the same and have just three possible structures as shown below:

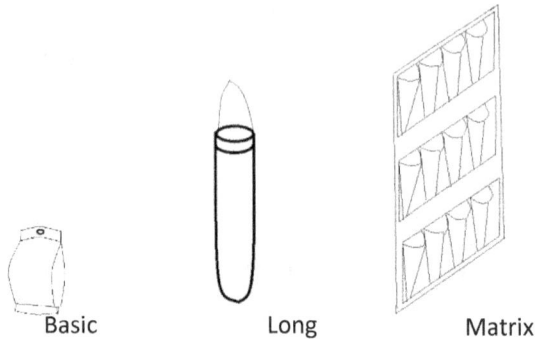

Basic   Long   Matrix

Fig. 7-3. Vertical Agriculture Growth Units

The VAGUs must fulfill the following requirements:
- Made from recycled materials (not reused)
- Located under any roof to protect them from heavy rain if there is not possibility of indoor environments.
- Used only for food.
- The growing medium or soil must be the same for all the plants.
- Minimum intervention.
- Minimum of resources.

Off course that there are some commercial units which can fit under this description, but it´s assumed that the following concepts are going to be as simple as possible.

7.6.1 Ethno units

These units are a low cost solution which can be implemented and used by people without major restrictions, because they should use the same elements as for traditional agriculture: soil, manual irrigation, nutrients and simple hanging accessories.

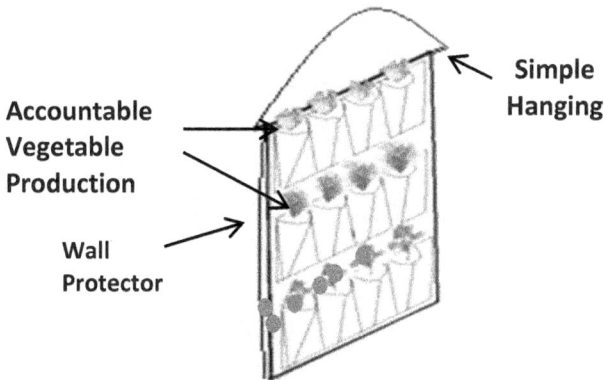

Fig. 7-4. Matrix VAGU Ethno.

No special materials are used to manufacture a VAGU ethno: Plastic or Polyethylene, rope, electric plastic sealing tool and scissors. The units are going to be hanging or supported on the wall. In this last case a plastic will protect the wall from damage because of any water leakage or moisture.

7.6.2 Techno units:

Since these units can include sophisticated control elements, the required knowledge for development can be high and the

Growing food in restricted environments

costs, depending of the level of technology can be very expensive.

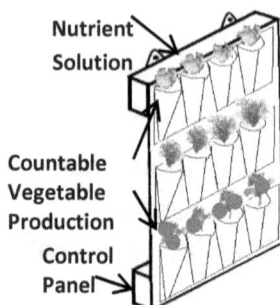

Fig. 7-5. Matrix VAGU Techno.

Techno units can use both systems to support the plant: Soil with nutrients, any type of substrate or not soil at all. If this last one is the option selected, then it´s necessary to develop the support of the plant for the whole growth cycle.

More technology is allowed here: Lighting can be natural and artificial, irrigation can be fully automatic, different sensors can be added for better quality control of production: ph, rh, temperature, etc.

The ideal VAGU Techno will lead to something amazing that can change the future. Envision a person of 2050, who lives in a small yet comfortable apartment in a very crowded city, who has to travel to another country for three months. Before departing for his trip he set the parameters of his VAGU techno, pushes the start button and leaves the apartment. The control panel on the VAGU will have control of the plants during the growth cycle and later, after three months when the person returns, will have the crop ready for harvesting and consumption. This is the expected future!

Growing food in restricted environments

7.7 Advantages of VAGUS

First of all, the VAGUs represent a simple, low cost and defined way to grow food. It´s possible to create different VAGUs for specific plants and crops since this is the purpose of this proposal: People have to develop growth units according to their own needs.

Regarding extended production with VAGUs there are even more advantages :

- Easier accountability/feedback of production
- Best data management
- Control of resources
- Less waste
- More possibility of successful crops
- Lower production costs
- Portable land and mobile crops
- Ready for space era
- Faster cultural migration
- Less agrichemicals
- More personal crops
- More interest on trading than commercial acts
- Standardization of resources
- Less soil degradation
- Less space occupied
- Less maintenance and care
- Better taste and nutrition
- Best human adaptation to climate change
- Materials are reused

Vertical Agriculture

Low cost Vertical Agriculture Growing Units with a nutritional/survival kit of vegetables to help small farmers with their family livelihood strategy and better sustainability.

7.8 Construction of VAGUs

Again, an important remark is related to the construction and materials of VAGUs is that these designs are only a suggestion and that can be changed, adapted or enhanced according to specific needs of plants.
The main objective here is to make it simple, affordable and with no much tooling.

7.8.1 Vagu Basic

This VAGU looks like a small pillow. Tough you can buy or re-use plastic bags; it is a good exercise to learn how to manufacture them.
The following pictures show the way they can be made. An important remark is the idea of adding the cover. The purpose for this is to help in preventing evapo-transpiration when the environment is hot, saving the water for less irrigation. Since this sealing retains the water it´s important to know about the water requirement of the plant to avoid any damage to the plant 'roots.

1. Select the size. 2. Seal one end. 3. Cover for the other

Growing food in restricted environments

4. Align the two parts.

5. Seal the other end

6. Open the interior

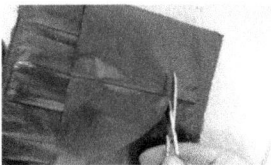
7. Open the interior

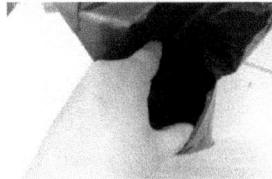
8. Open the interior

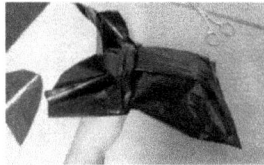
9. Fill it with soil

Fig 7-6 Making a VAGU Basic

7.8.2 VAGU Long

This system is similar to some experiences with strawberries made on long plastic tubes or PVC tubes. A small difference is the way to hanging, the total size and the possibility to "transform" it to a VAGU Techno.

Size is important because the plants have to grow freely, the weight of soil makes the VAGU units too heavy once the plants have fruits, and the column of weight compacts the soil in lower levels so irrigation could become a problem.

Reusability is also important so, after a harvesting process, the soil can be removed and prepared back for a new production by simply un-tying a rope.

1. Get a long plastic tube

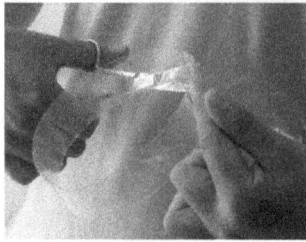

2. Cut a plastic strip

3. Put the plastic tube inside.

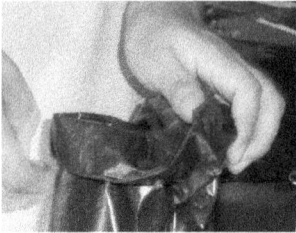

4. Fold several times

5. Tie the other end

Fig 7-7 Making a VAGU Long

7.8.3 VAGU Matrix

Every VAGU follows requirements of size and supported weight. The measurements and material must be decided according the intended crop, but it is certainly important make small and portable units. For experienced and industrial farmers, big size units can lead to better production and economic results, but off course, this will require some infra-structure.

The following pictures can be considered as simple suggestions since the VAGU can be constructed for specific purposes; however this standard sized has been tested under simple experiments as seen later on this book.

1. Define the VAGU size

2. Fold with some extra length

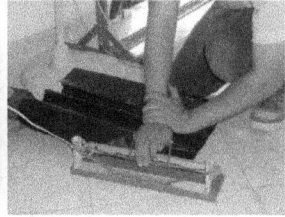

3. Fold and seal

4. Make a transversal sealing

5. Cut and separate the bags

Fig 7-8 Making a VAGU Matrix

7.9 Cases of study

The following experiences show the usage of Ethno Vagus in the production of different vegetables. The water irrigation has been manual and no chemicals have been used. The soil is normal or commercial soil for plants with some nutrients

7.9.1 Vagu Basic

Vertical Agriculture

| Day 1 | Day 10 | Day 15 | Day 20 | Day 21 |

| Day 22 | Day 23 | Day 24 | Day 33 | Day 34 |

| Day 37 | Day 42 | Day 46 | Day 47 | Day 49 |

Growing food in restricted environments

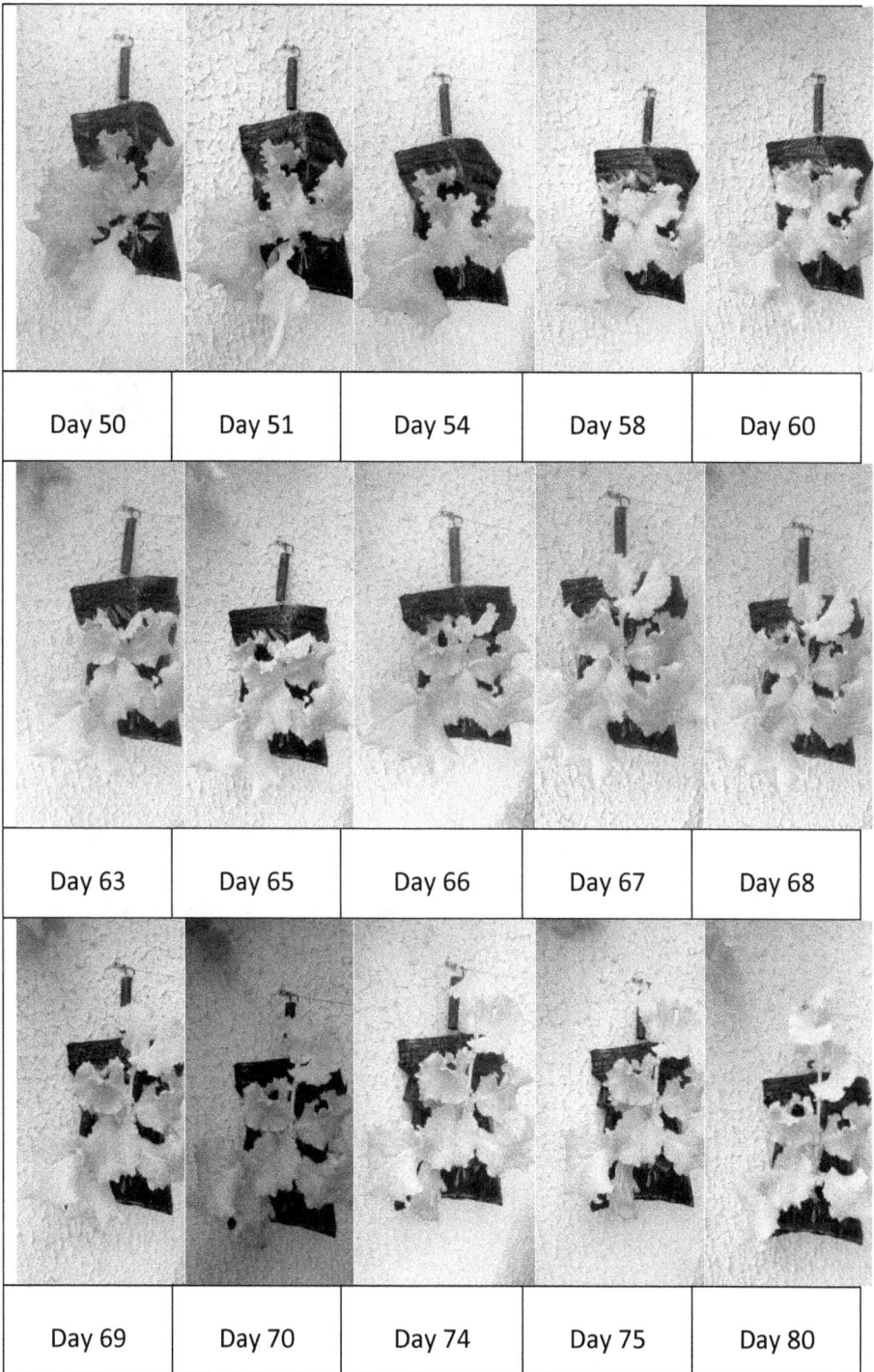

| Day 50 | Day 51 | Day 54 | Day 58 | Day 60 |

| Day 63 | Day 65 | Day 66 | Day 67 | Day 68 |

| Day 69 | Day 70 | Day 74 | Day 75 | Day 80 |

Growing food in restricted environments

Fig 7-9 VAGU Basic in Production

7.9.2 Vagu Matrix with good sunlight exposure

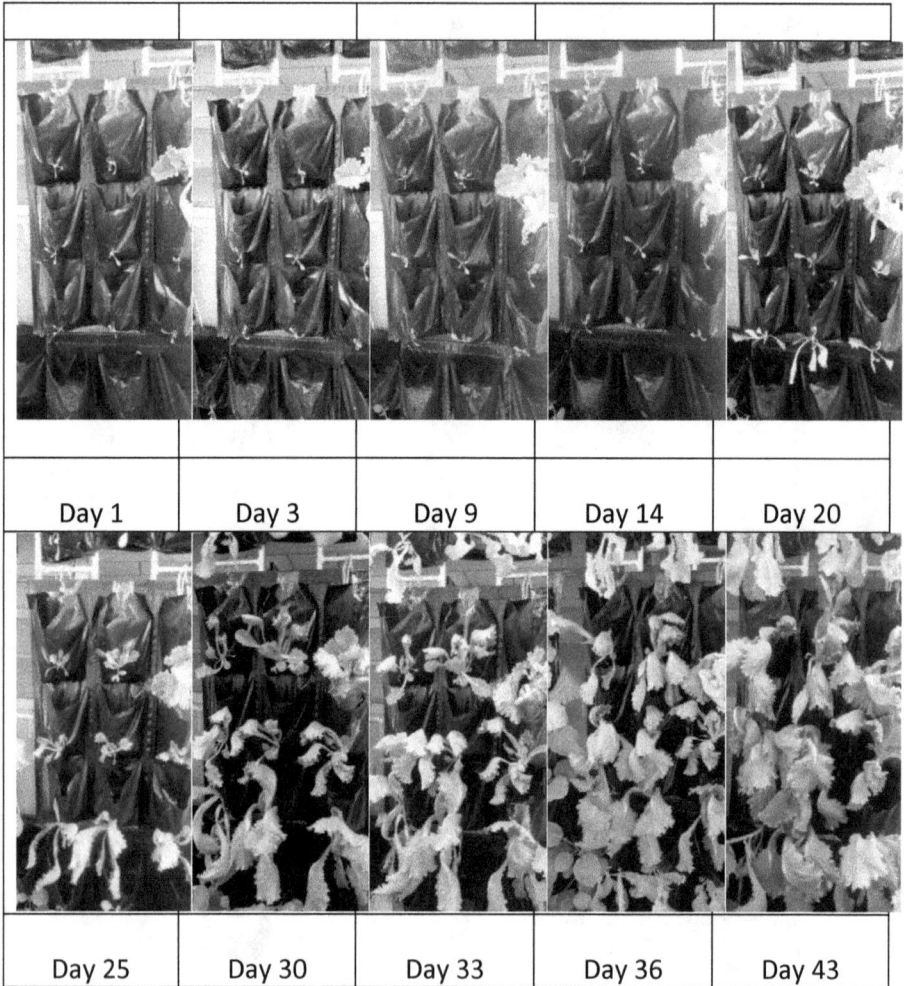

| Day 1 | Day 3 | Day 9 | Day 14 | Day 20 |
| Day 25 | Day 30 | Day 33 | Day 36 | Day 43 |

Fig 7-10 VAGU Matrix in Production

Growing food in restricted environments

7.9.3 Vagu Matrix with little or insufficient light exposure

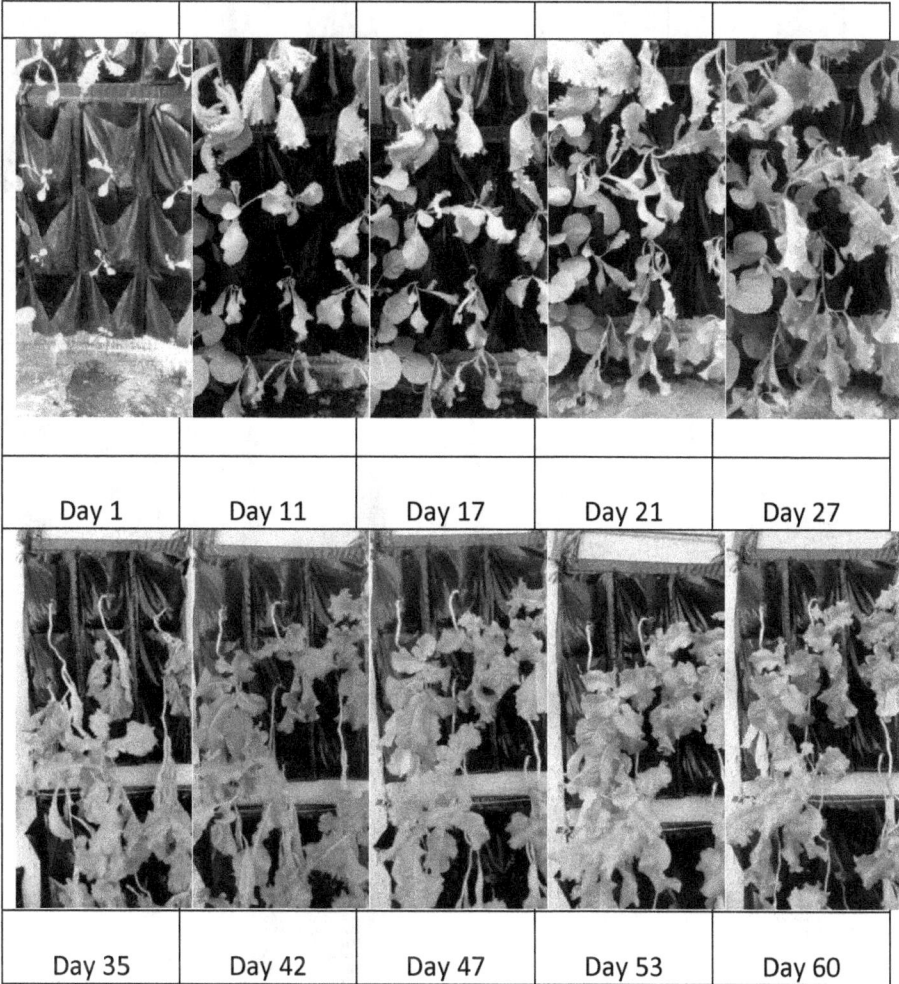

Day 1	Day 11	Day 17	Day 21	Day 27

Day 35	Day 42	Day 47	Day 53	Day 60

Fig 7-11 VAGU Matrix in Production (away from light source)

7.9.4 Vagu Long.

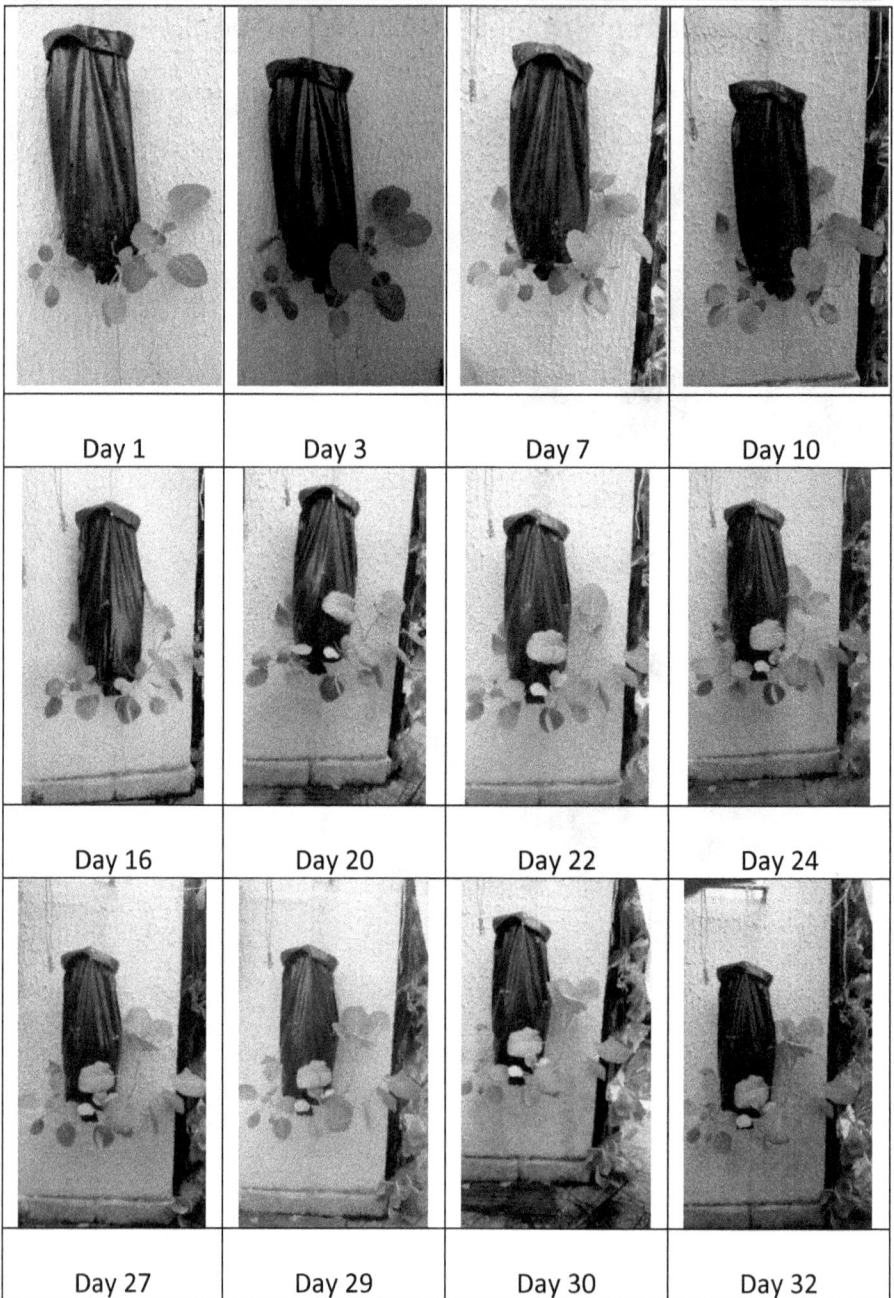

Day 1	Day 3	Day 7	Day 10
Day 16	Day 20	Day 22	Day 24
Day 27	Day 29	Day 30	Day 32

Fig 7-12 VAGU Long in Production

Growing food in restricted environments

7.9.5 Vagu Matrix

Day 1	Day 3	Day 8	Day 18
Day 23	Day 34	Day 41	Day 47

| Day 57 | Day 67 | Day 75 | Day 79 |

Fig 7-13 VAGU Matrix with another edible vegetable

7.9.6 Vagu Techno

Currently, this type of Vagu is under development. Some units have been tested but it is clear that, because of its numerous advantages, the Vagu techno is more expensive and requires more components.

One of the most difficult tasks in designing the Vagu is to have a low cost unit without compromising the total reliability, because the purpose of Vagu techno should work almost fully unattended.

Consider the production of tomatoes. This is one of the most popular fruits (yes, the tomatoe is a fruit and not a vegetable).

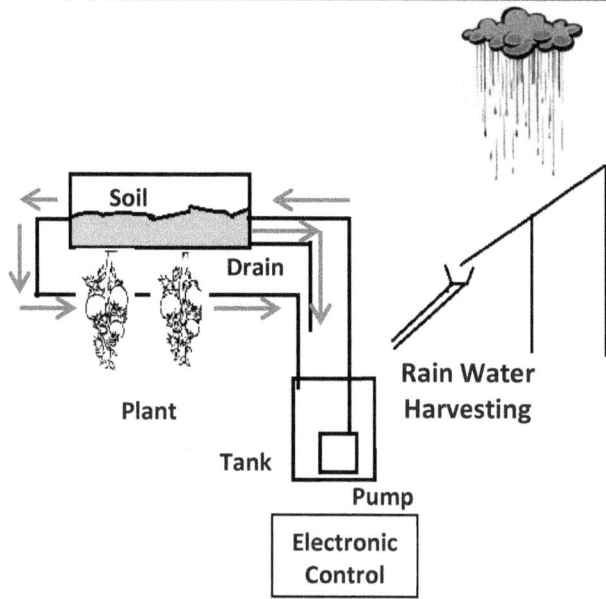

Fig 7-14 VAGU Techno for Tomatoe's production

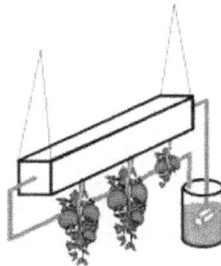

Fig 7-15 VAGU Techno for Inverted tomatoe production

The unit can be irrigated manually or through an automatic control box which takes care of the solution with nutrients to feed the roots

Fig 7-16 Different images of inverted tomatoe production

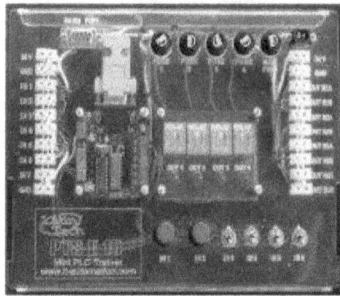

Fig 7-17. Control system with PLC trainer

7.10 The stages to grow food in VAGUs.

There are several simple stages:

1. Select the seeds and define the maximum plant growth.
2. Plant the seeds on a seedbed.
3. Transplant to the VAGUs.
4. Make periodical irrigations.
5. Check for normal growth.
6. Remove undesired species or resolve minor problems.
7. If ready go back to step 8. If not go to step 4.
8. Collect your product.
9. Document your acquired knowledge and share.

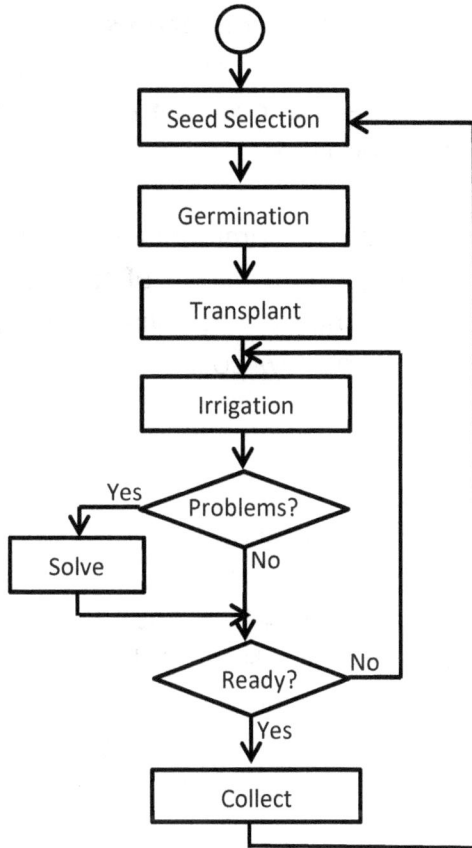

Fig 7-18 Flow diagram to learn in vegetable growth.

7.11 Why this project can be successful?

Several projects have shown that empowering small farmers to learn and grow food under new techniques could lead to a new and more efficient way to produce food.

The construction of knowledge (initially for the Ethno units) will be collective and participative. The small farmers could grow some selected and nutritional vegetables; use them in their daily diets to improve human and specially children´s health. Since the plants are not in contact with soil, there is less need

Growing food in restricted environments

for herbicides or pesticides. Since the VAGU´s are located on external walls and under roofs, the weather (rain or droughts) won´t have direct effect on them. The maintenance is very little and only manual (but could be automated) water irrigation (with nutrients) is required.

Groups of individuals or communities, according to their own need, could grow different types of food and then generate their own trade economy, improving their health and living conditions. With some low cost and simple control we could avoid any water waste.

7.12 How can it be scaled up?
People (fans or farmers) can be trained to grow food with VAGU tailoring the education and units according to their specific needs (ethno and techno). For instance, kits could be prepared in two ways: containing everything (seeds, nutrients, controls, and hangers) or simply the vertical units to grow whatever they want as long as they are familiar with their plant size restrictions.

Any country interested in helping a given population suffering hunger, could send a lot of survival VAGU kits even with the needed water supply to guarantee full production. If the kits go on board a ship, vegetables can grow during the trip.

Hypothetically speaking, succeeding in having small scale food production new promising perspectives arise. People and small farmers could grow some selected and nutritional vegetables; use them in their daily diets to improve human and specially children´s health. Since the plants are not in contact with soil, there is less need for agrochemicals (such are herbicides or pesticides) to protect the small crop. Groups

Growing food in restricted environments

of individuals or communities, according to their own need, could grow different types of food and then generate their own trade economy, improving their health and living conditions. With some low cost and simple control we could avoid any water waste.

VAGUs can be designed according to specific needs (with or without control or automation elements, using simple tools with normal plastic sealers, but it needed the process can be automated for mass production looking for lower manufacturing costs .

7.13 Methodology.

The initial stage is to interview people, mostly women because of their key role on family nutrition, in urban and rural areas to know more about the way they grow food, the perception they have of traditional agriculture and the interest on getting a new knowledge. The research will allow understanding of their needs, the kind of situation they have and the way a vertical grow unit could affect their future life. A preliminary Vertical Agriculture transitional model will be formulated by using Soft System Methodology (from P. Checkland) which uses a systemic approach to complex problems.

For initial testing, different groups or individuals should be introduced to use simple VAGUs to produce food, after a simple training and then learn from the experience. The groups and individual can be categorized by location, experience, background, interests, importance and problems.

The initial stage is to interview people in urban and rural areas to know more about the way they grow food, the role of women and men in the food production, the perception they have of traditional agriculture and the inception of a new knowledge.

For instance a community of families with a low income, not healthy living conditions and with experience on growing food outdoors and with traditional ways, could be a very interesting group to measure the experience.

7.16 What is the next step?

Though the first experiences of VAGUs have been positive and proven to be a low cost alternative, a more extended use would be beneficial. It is encouraged to develop more research on this field with the following focus:

- Development of new VAGUS. There are several plants and shapes so it´s required creating a proper VAGU for each of them, according to weight, space, location, resources, etc. The more VAGUs the lower the cost.
- People without any restriction should try to experiment to grow food by themselves: from simple vertical gardens in homes, small farmers, communities that want to improve food availability, up to industrial growers which see an important market.
- Study of population and the way they embrace, reject, support or share any acquired knowledge. This work becomes important to provide the needed feedback to make changes on any methodology.
- Categorize population in order to make more suitable applications based on real needs. For instance, some of them would prefer to work on medicinal herbs, yet edible species, rather than nutritious food.
- Encourage nutrition as the main focus. This will help in population health and economy.

Growing food in restricted environments

7.17 Where Vertical Agriculture can be encouraged or applied NOW?

This is a vision for the future, but there are some cases where there is an urgent need:

- Places with extreme weather events or with awareness of climate change effects.
- Communities or groups with related interests.
- Isles or locations with restricted resources.
- Countries with extreme poverty.
- Countries with internal problems like wars, weak economies or difficult governments.
- International help for easy accountability. Less demanding labor and lower production risks.

7.18 Is it applicable to all type of plants?

Many economies of the world are based on crops like Coffee, Corn or Sugarcane. For centuries, they have been typically cultivated using the classic techniques of horizontal agriculture but under this new vision, there is a big doubt if it´s possible to do it with Vertical Agriculture. The answer is that it´s really possible to develop the vertical growth units to carry this type of plants.

It´s important to understand that there are no major limits to grow any type of food as long as the design of VAGUs tries to fulfill most of the golden rules.

The main problem when defining the VAGU for any vegetable species is related to some factors:

- Weight: The VAGU must be designed to support the weight from a single to many units of vegetables in adult state and with fruits.
- Height: The maximum height the plant requires to grow healthy and also the distance from other species to allow the correct growth without interference.
- Water/ Nutrient requirements. The VAGU must be able to offer equal amounts of the solution for any specie in the VAGUs.
- Natural/artificial light exposure. The plants must have enough exposure to the sunlight to grow healthy and preventing that long or big leaves can block other plants in the same VAGU

When the space to grow food is restricted, one of the main problems to handle is precisely the efficiency of the crop. The main idea is to have food but some species have to grow branches and leaves before they can have fruits or seeds.

The ideal is to have species with nutritional content which can grow as short as possible, without decreasing its possible yields. Several questions arise: Could we use techniques to manipulate the size and the efficiency of the species? Is it important to grow shorter versions of the vegetable we know now? Will be they accepted by population?

NASA and Utah State University have been testing a super dwarf wheat which can be grown in space and off course because of the space limitations of a spaceship. The first variety was (USS-Apogee) of 40 cm, which has a height much smaller than the normal wheat (around 100 cm). Recently another variety called USU-Perigee has been developed and it can be grown in small spaces because it reaches a short height of around 30 cm.

Growing food in restricted environments

Fig 7-19 Dwarf Crops for Space Flight.
Source:Crop Physiology Laboratory. Utah State University 2015

7.19 Golden Rules for Vertical Agriculture

Since agriculture is a wide concept, it is important to have some preliminary rules to help on the implementation of Vertical Agriculture projects. These simple rules are as follows:

- **Between ornamental and edible plants, Vertical Agriculture growers must choose always food.**

Our future depends on food so why spend money and resources on vegetable beauty when you can use your energy and resources to grow food, which can also be beautiful. Even if there is no interest on consumption, growing food will help us to learn, share and have a social impact.

- **Depending on the environment, the crop must be under a basic roof (natural or artificial) as minimum requirement, never fully exposed.**

Growing food in restricted environments

Depending on the location, the size and the different climates the crop will experience throughout the year, the requirement will go either from a simple roof or cover up to a complete greenhouse. Even if there are no major risks for the crop, it is convenient to start protecting the crop. Rains, droughts, storms, extreme heat, freezing etc. will be part of the new weather because of climate change.

- **Every project must be made on recycled (not re-used) materials.**

Please take into consideration what is explained about plastics in other chapter. Don´t re-use don´t help the plastic manufacturers to bring more plastic to the environment, better send them back all the plastic you can for them to mix it with their current production. Even if this is a new product, all plastic should return to the manufacturers or source.

Although in some cases re-use of plastic can be allowed, when the amount is minimum (like one to ten relation) in general try to avoid it.

- **The design of the system must consider the minimum of input resources**.

Not only economic considerations. Using less materials and optimizing the design in weight, size, portability, effectiveness will help to spread the knowledge and if, possible, the technology.

- **All projects must fulfill the criteria for maximum and minimum amounts**.

Every aspect of growing food must be carefully analyzed: Nutritional content, maximum growth, harvesting time, production costs, environmental impact, material and production waste.

- **Growth and evolution of vegetable species must be as natural as possible.**

Artificial lighting and modifications of natural cycles are allowed if there is no other option. Hybrid solutions are encouraged.

- **The design must be as ecosystemic as possible and must propend for diversity.**

Since the intention of Vertical Agriculture is to avoid an possible damage to nature in all its expressions, all cares must be taken to prevent soil degradation, waste of resources like water and nutrients. If the crop is going to be along with other species as plants, the growth must be without any invasive aspects.

- **Trade and not commercialization must be encouraged.**

Off course commercialization of food is allowed, since this is the focus of most of the current agriculture, but it´s expected that people become interested in producing their own food under VA concepts and it would be great if people use it as a tool to learn about community concepts and the power of sharing.

- **Only what can represent a risk to the production must be controlled.**

Yes, production can be fully organic, but sometimes it´s require eradicating of problems which can surge because of contamination and wrong practices. In this cases,

Growing food in restricted environments

agrichemicals can be allowed but as a specific case and not as a common practice.

- **Type and Handling of waste must be estimated prior to an AV implementation.**

The ideal situation is not to have any waste of food or installation products, but in case that you could have some wastage or residual material as a result of your operation, all consideration must be studied to prevent possible pollution situations. The acts must be responsible.

- **Any knowledge acquired must be shared.**

Vertical Agriculture must be from people and for people. This approach is a re-discovery of processes and techniques. Sharing of all your knowledge is encouraged and mainly because we are living a communication era where there are no limits for information and there are no delays in getting it.

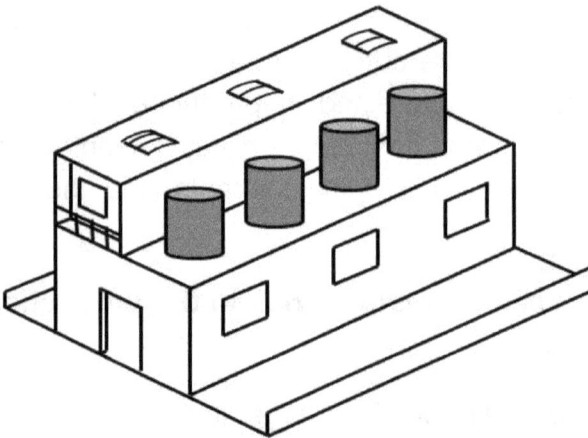

Growing food in restricted environments

References:

PLC trainer and automation
http://www.lt-automation.com/PLCtrainers.htm

Checkland Peter, Soft System Methodology.

Dwarf Crops for space flights
http://cpl.usu.edu/

Chapter 8

Climate Change.

"Climate change is happening big deal here in the Arctic. And it is our decision to trying to change this. So: let's do something about the biggest threat of our time. Maybe we cannot save this bear here. But every little action we do to change our ways is a step in the right direction. We just have to get started and keep on going! "

Kerstin Langenberger Photographer

Fig 8-1. Freeze in Bogotá,
Colombia. Source: El TIEMPO
(30/03/15).

The NASA from USA has recognized the evidence that global warming is a man-made condition. Sea level and acidification are increasing and that's a cruel truth. The temperature of the planet is also rising and the glacial and perpetual snows are shrinking specially in the artic sea. Extreme event are appearing more often.

The IPCC report of 2014, based on a collaborative job from around 2500 researchers form more than 130 countries, also concluded that the climate changes has an anthropogenic origin.

The basic reason is that the population is producing more carbonic gas than plants and ocean can absorb. In addition, the gases remain in the atmosphere for years even if we succeed in removing the harmful ones we have today.

Some scientists explain that cycles of cooling and warming are a phenomenon which can happen after several thousands of years, as a product of slight changes on the terrestrial orbit and a different exposition of the planet to the sunlight.

Possible effects because of climate change

- Changing patterns of infectious diseases.
- Extreme weather and winds (tornadoes, floods, hurricanes storms)
- Fluctuations on short duration temperature events
- More presence of pollen and other aerial allergenics because of high temperatures.
- Extreme temperatures like very hot days and very cold nights.
- More frequent rains and more long duration droughts periods.
- Compromise of water availability
- Heavy rains will affect agriculture and food availability
- Diseases like diarrhea and malaria because of water quality.
- Risk of places because of avalanches of land and water.

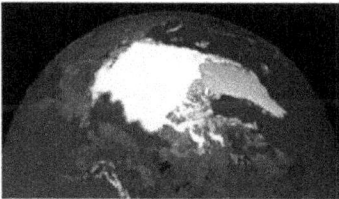

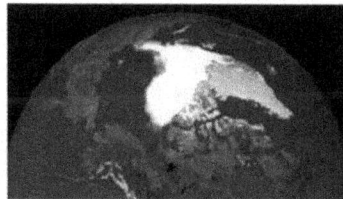

1979   2007
Fig 8-2. Source: NASA (2007). Sea Ice Yearly Minimum 1979-2007. These two images, constructed from satellite   data, compare arctic sea ice concentrations in september of 1979 and 2007 (Images country of NASA

Growing food in restricted environments

The planet is getting warmer. According to NASA´s Goddard Institute for Space Studies (GISS), the warmest average temperature of the planet, analyzed from 1880, was reached in 2011. The eight warmest ones occurred since year 2000.

This institute has been monitoring the global surfaces temperatures and compares them to the ones form the midht-20th century. The results clearly show that there are higher temperatures than former decades, for instance, in 2011 the temperature was 0.5°c higher than the mean.

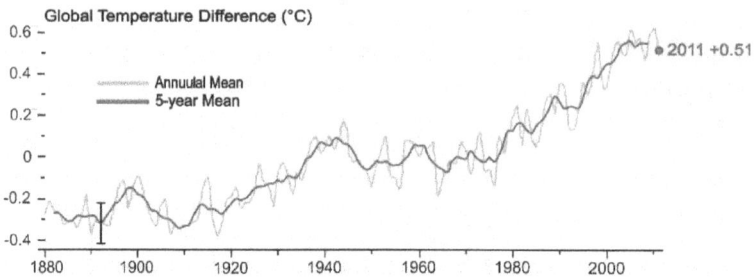

Global Temperature Difference (°C)

Annuual Mean
5-year Mean

2011 +0.51

Fig 8-3 Data source: NASA Goddard Institute for Space Studies. Image credit: NASA Earth Observatory, Robert Simmon

The planet is experiencing a trend to increase the average temperature over the years, as shown on the above graphics. The causes can be different and anthropogenic, but effects are already well known.

Fig-8-4 Source: Kerstin Langenberger Photography . Svalbard. Norway.

Growing food in restricted environments

Resources:

Kerstin Langenberger
http://www.arctic-dreams.com/

Snow in Bogota
http://www.eltiempo.com/bogota/clima-en-bogota-viviendas-afectadas-por-granizo-en-el-sur/15447978

Climate change and global warming
http://climate.nasa.gov/evidence/

IPCC 2014 climate report
https://www.ipcc.ch/report/ar5/

Sea ice
http//svs.qsfc.nasa.gov/vis/a000000/a003400/a003464/index.html.

Warmest Temperature of the planet
http://data.giss.nasa.gov/gistemp/

Chapter 9

Food security and Sustainability.

"Saving our planet, lifting people out of poverty, advancing economic growth... these are one and the same fight. We must connect the dots between climate change, water scarcity, energy shortages, global health, food security and women's empowerment. Solutions to one problem must be solutions for all."

Ban Ki-moon

The best way to understand about how to live sustainably is finding the way our lifestyle doesn't have negative effects on environment and other people. The problem is that as population grows there is more stress on available resources and also it generates a competition for them.

More precisely, sustainable development (term extracted from the paper Our Common Future, from the Brundtland report or more recognized as the World Commission on Environment and Development (WCED)) is defined as the development that looks and meets the needs of the present without compromising the ability of future generations to meet their own needs. In general terms, it is the idea of envisioning the future based on what we do today by avoiding all the possible mistakes which can cause problems to future generations.

According to New Millennium Goals, in order to achieve sustainability for a growing population, there is urgency in developing new agriculture methodologies. Under the new conditions of climate change and lack of arable land the possibilities, for the whole population, mainly small farmers, are scarce since growing food outdoors can be risky and requires a lot of resources to succeed.

People with no land will require producing some food with minimum resources or care, this condition includes not only small farmers and rural habitants but population living in urban centers.

But how eradicate hunger in present times? FAO has release what is called the latest's hunger map, where you can see that African and Asian countries show a predominant rate of undernourishing. In Latin America, although the problem

Growing food in restricted environments

seems to be moderately low in the whole region, some countries are facing a worsening situation.

This map is not considering the risk for food production under the climate scenarios of floods and droughts, the use of polluted waters and other local situation like wars or earthquakes.

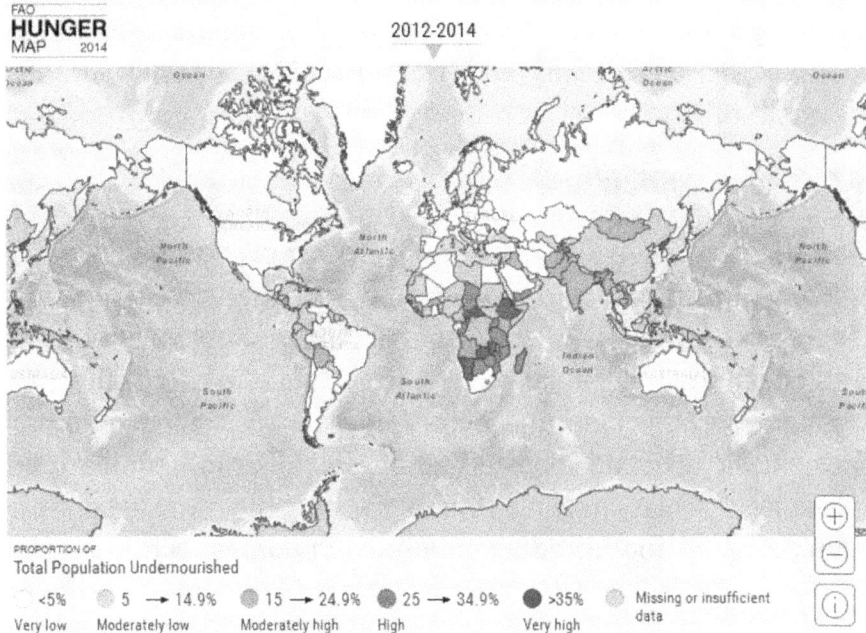

Fig 9-1. Hunger MAP. Source FAO 2014

The united nations has launched a program called the Zero Hunger Challenge, as the next goal to succeed on avoiding hunger for 800 millions of people.

Fig 9-2. Zero Hunger Challenge. Source UN 2014

Growing food in restricted environments

Six focuses are already identified:

Nutrition guaranteed in order to have stunted children less than 2 years, ensuring of full access to adequate food all year round, encouraging of all possible actors to make all food systems sustainable, helping the increase in smallholder to increase his productivity and income and minimization of loss or waste of food.

9.1 Role of Agriculture

The key to assure nutritious food for a growing population is the agriculture. Not as a business but as a method.
There is a huge need to embrace and promote agricultural practices that avoid the use of agrichemicals such as pesticides, fungicides and even fertilizers. Focusing on small farmers, it is quite possible that soil degradation can be prevented while helping them to become a sustainable alternative for food production which competes with industrial farms.

Sustainable agriculture is sometimes misunderstood as "organic" agriculture.
If the food has labels like "100% organic" or "Made with organic ingredients" it mainly means that it was grown without any synthetic agrichemical (pesticide, fungicide or fertilizer). It also implies that, although it is difficult to control or verify, not other chemicals like antibiotics and growth hormones have been used.

In some countries, the sewage sludge is used as fertilizer, carrying a lot of dangerous chemicals which can affect human health.

Organic food production has been one of the fastest growing sectors of world agriculture. Working organic products is often more profitable per acre than conventional agriculture production since farm inputs to get higher yields imply more costs since costs associated (workers, fuels, fertilizers, pesticides, tractors, storage) used in conventional farming are bigger. Some research found that organic production could receive half of the nutrients compared to conventional agriculture.

Sustainable agriculture must be based on the correct understanding and respect for the environment and the biodiversity, trying to be responsible on any possible action which can generate harmful effects in present or future.

9.2 Is it possible to think differently?

One of the most difficult tasks is to convince people to think and act different from the normal way they have been growing food.
Traditional agriculture growers are used to associate the need of soil as the only way to develop vegetable cultures. Considering a soilless agriculture could lead to problems because of the ancestral knowledge can't be questioned or irresponsibly discrediting its importance.

Assume that small farmers grow tomatoes. They normally will do it on the ground and based on simple irrigation as show below.

Fig 9-3. Irrigation of plants on the soil

One of the problems is that the culture has to fight against other species like herbs, survive to bugs and also the water evaporates because of the sun exposure. To avoid these problems, some improvement is achieved by using plastic beds like the ones you see on the picture.

Fig 9-4. Growing strawberries on plastic beds

If the agricultural practice is made outdoors then there is more vulnerability to cold or heat. This situation can be overcome with simple cover or protection techniques like the one show below:

Growing food in restricted environments

Fig 9-5. Low cost greenhouses

Now, because most of the diseases of the plants come from the soil, imagine a structure that can be hung on any internal home wall and that can grow different vegetables without much care and without any effect of climate conditions.

Resources:

Hunger map
http://www.fao.org/hunger/en/

Zero Hunger Challenge
http://www.un.org/en/zerohunger/#&panel1-1

Chapter 10

Plant Growth and Paradigms.

"Out beyond ideas of Wrongdoing and Rightdoing, there is a field. I´ll meet you there".

Jalal Ad-Din Rumi.

Vertical Agriculture

Nature always finds its ways to develop. For several centuries the growth of plants has been a focus of study, more than any other topic, in order to understand how population could obtain successful crops under more adverse environments.

Knowledge on plant growth can be redefined with the research or with the experience:

- **Vegetables (food) can´t grow without care.**

Though with care plants can have a better growth, several species of plants manage to survive alone in wild environments. These species require little care other and grow in distant places where there is no possibility for any human intervention. Good soil, water availability, some shade and proper temperature are minimum requirements for simple, not necessarily high quality, plant growth.

- **Small farmers should continue with traditional agricultural technologies**.

One of the main mistakes is trying to keep traditions alive no matter if they are not that efficient. It´s not a matter of changing the traditional knowledge by applying technology at once but helping to any community to make a smooth transition with cultural transformations toward a better and more efficient way to produce food. Since these proposed cultural transformations could take life generations, there is a big chance of having people to develop new technologies and knowledge by themselves.

Should we continue using the same agricultural practices to satisfy the needs of food for a growing population?

Growing food in restricted environments

- **It's the same to plant in vertical than horizontal.**

Traditional or conventional agriculture has been mostly made locating crops on a horizontal way and because of this practice most of the usage of space is extensive. Technologies and mechanized equipment have been developed around horizontal crops or better understood as one-level crops.

Though most the knowledge acquired and developed with conventional agriculture can be used on vertical crops, the proposed methodologies for vertical agriculture will require either adaptation or the development of practical and innovative concepts.

- **Artificial light is not good for plants.**

There are a lot of technical discussions regarding the sunlight and the benefits they have on plants and photosynthetic processes. The truth is that, beyond any possible speculation, there is a huge interest in growing plants indoors, so the scaling of usage and availability of technologies on artificial light will bring costs down and will make the artificial light as good as the natural light. It's just a matter of time.

- **Plants will adapt by themselves to the new climate change scenarios.**

Adaptation of plants to extreme environments will clearly happen because of the natural of the species on surviving to cold, hot and rainy weather exposure.

This process, however, could take years and it would make it difficult to the sustainability to wait for a successful process while the population is starving. Artificial training can

Growing food in restricted environments

accelerate the process through the use of climate chambers also known as Phytotrons.

- **After extreme environmental events the plants recover their original condition.**

When plants experience extreme conditions, for example the freezes in Brazil, they don't recover easily nor continue on the same production level. Temperature can affect photosynthetic enzymes, electron transport on chlorophyll and apply stress on the leaf, changing the evapotranspiration rates. In general plants have effects on metabolism and biomass production.

- **How adaptation should be? Nature to plants or Plants to nature.**

This is a difficult question. Plants require surviving in future hostile environments like extreme cold or heat as a consequence of climate change. In addition to controlled environment applications from indoor cultivation, which is a way of adaptation, a parallel way has to be developed on the species for them to be more resilient when exposed to extreme weather. This research on new seed, test and training on plants, could take many years while the availability of food is becoming a more serious situation every time.

A faster solution is based on developing proper growing environments, preventing any stress or harm, for every plant so all the conditions are met during their life-cycle. This goal is certainly not easy nor that affordable.

- **What is the right environment for plants?**

Every species is different and depends on the way you want to have the crop development. Crops with single species are a good idea when you are just beginning experimentation. Resembling wild environments could be a little different since a diverse group of species will have different requirements and will have different effects on the others. Some vegetables require exposure to specific patterns of humidity and light.

Regarding to the possibility of having a crop with diverse types of vegetables, sharing a common place, it is just a matter of investment and research to develop programmable climate zones with specific consumption of resources, dispensed at different but programmable rates.

- **Should agriculture be made just horizontally and one level?**

The more we learn on saving space on agriculture, the more efficient use of resources we make so, it is worth to try to take advantage of any location, no matter how small or big it is. Conventional agriculture wastes space, because of the few application restriction it has. Having enough vegetable production close to consumers can save a lot of economic and environmental costs. Improvements on vertical agriculture could generate a new and more efficient way for food production.

- **Are the pesticides and some other agrichemicals really necessary?**

They are mostly needed when the crop is outdoors and there is no much control on the environmental parameters and the

Growing food in restricted environments

pathogen agents such as insects, fungi and invasive species such as herbs.

It´s supposed that when the indoor crops have control on the environments, for example, cleaning or sterilization practices; there is no much chance for pest, fungi or herbs to grow along.
Some cases will require using of some agrichemicals but it could be barely used and applied at minimum.

- **Vegetables must be produced in favorable environment places.**

This is something to be learned by trial and error, but with a simple roof, the VAGUs and the design to avoid losses of humidity, there are more chances for any plant to survive in any hostile environment. The plant will do the rest and it will accomplish the necessary to adapt to new climate conditions.

- **Food must be grown on a large scale**

It is worth to try to have individuals growing their own vegetables. This excess of food can be an interesting test of sustainability. Having people growing small and diverse amounts of vegetables will produce unprecedented effects on agriculture, markets, health, nutrition and culture.

- **Only specific species can be planted indoors**
Any plant can grow indoors as long as you provide the required light, water and nutrients. Off course, growing plants indoors could require artificial supply of what the plant is lacking and can obtain growing outdoors. High investments could be required to achieve this kind production.

Growing food in restricted environments

- **Technological solutions are very expensive and not available.**

The only problem is that applications need to become popular and the costs will go down. In automation and control for artificial environments mostly everything can be made as space and scientific research has been proving all over the years. For Vertical Agriculture, technology is not a specific requirement.

- **Population and mainly small farmers must buy the technology and not developing it by themselves.**

This is something that needs to be solved. Making low cost solutions for automated vegetable production is possible, as long as you use generic products and not brand oriented devices and integrations. It is important for people develop knowledge by themselves and share with others what they have learned.

- **We should continue our agricultural practices outdoors.**

No, the climate change is having more harmful effects on agriculture, because rain, droughts and heat are increasing. If we continue on the same path, with the se same agriculture, there is a big chance of losing everything.

- **Plants and nature will adapt to new scenarios of climate change.**

No, we can't wait that long. It´s possible that this process take years and the need for food is now. Better learn how produce food in environments with restricted space and resources. This is precisely this proposal of Vertical agriculture. There are many factors affecting the plant growth: weather, light including heat, temperature, relative humidity (RH) , airflow,

Growing food in restricted environments

carbon dioxide (CO2), water, nutrients, plant pattern distribution and location.

One of the companies which is working on different and innovative ways of food production is the Dutch company PlantLab *.

PlantLab has researched over the best combination of factors which are appropriate for a given species of plant and, under a technological conception, a so called "paradise" is prepared for the life cycle of the plant to guarantee an enhanced and trouble free growth.

Before any practical application, a mathematical model which describes the plant growth has to be developed.

The company claims that under these controlled environments, the crops will grow consistently, fresh, free of pesticides and without any adverse effects from climate, location, season or time of the day. Not to mention, that there is a promise for better taste and enhanced nutritional values.

A final question has to be added here: what do we want? Starvation because we protect the vegetable species or should we secure food for population no matter the risks this decision involves?

Growing food in restricted environments

Resources:

Paradigms on plant growth.
http://post.queensu.ca/~biol953/Casper Christiansen -The opposing paradigms in resource limitation on plant growth.pdf

Morris CE, Bardin M, Kinkel LL, Moury B, Nicot PC, et al. (2009) Expanding the Paradigms of Plant Pathogen Life History and Evolution of Parasitic Fitness beyond Agricultural Boundaries. PLoS Pathog 5(12):e1000693. doi:10.1371/journal.ppat.1000693
http://cedarcreek.umn.edu/biblio/fulltext/PLoS_Pathogens_2009_Mo rris.pdf

Plant Lab
http://www.plantlab.nl/

Chapter 11

Nutritional Facts of Plants.

"All plants are our brothers and sisters. They talk to us and if we listen, we can hear them".

Arapaho. American tribe

There are four clinically proven indicators for human health. They are associated with the kind of food available for consumption, but that can be affected by nutritious vegetables.

Insulin levels, Weight, Blood pressure and Cholesterol levels.

Most of the developed countries suffer of these diseases because of the food availability and the possibility to buy almost any food. This situation is quite different in developing countries, where several combined factors act against the population producing assorted effects from starvation up to nutrition problems.

11.1 Facts about nutrition

In general, it is necessary to grow healthy and nutritious food which should help population to make a sustainable living.

The less agrichemical involved the less harmful effect on the people. Several researches have pointed out that most of the death causes are originated by the chemicals used in agriculture, to grow plants protected from invasive species and harsh environments.

In addition to this problem, the water used for irrigation can be easily contaminated by other external industrial processes like mining, chemical spills, human wastes or deficient treatment.

The Environmental Protection Agency (EPA) considers the available herbicides (60%), fungicides (90%), and insecticides (30%) to be carcinogenic.

Beyond the former discussion and assuming that chemicals are controlled by the technology, it is worthy to focus on the kind of vegetable species which can guarantee the better nutrition per intake. In future days, availability of food can be at risk so , it is necessary to make sure that the portion of food administered to any human being contains most the needed elements for a healthy growth and with the best assimilation rates. The reasons to grow the best nutritious vegetables overcome the indoor or outdoor agricultural limits.

The U.S. Department of Agriculture National Nutrient Database allows the understanding that the content of Vitamins K and C, Lutein, Potassium, Fiber content and the amount of calories, could be good criteria to choose the vegetables to grow. The good nutritional profile makes the following vegetables as the best option for agricultural practices with the intention to enhance nutrition in future generations so it is important to read it carefully.

Vertical Agriculture

Nutrition Facts

Raw, edible weight portion.
Percent Daily Values (%DV) are based on a 2,000 calorie diet.

Vegetables (Serving Size)	Calories	Calories from Fat	Total Fat (g / %DV)	Sodium (mg / %DV)	Potassium (mg / %DV)	Total Carbohydrate (g / %DV)	Dietary Fiber (g / %DV)	Sugars (g)	Protein (g)	Vitamin A %DV	Vitamin C %DV	Calcium %DV	Iron %DV
Asparagus	20	0	0 / 0	0 / 0	230 / 7	4 / 1	2 / 8	2g	2g	10%	15%	2%	2%
Bell Pepper	25	0	0 / 0	40 / 2	220 / 6	6 / 2	2 / 8	4g	1g	4%	190%	2%	4%
Broccoli	45	0	0.5 / 1	80 / 3	460 / 13	8 / 3	3 / 12	2g	4g	6%	220%	6%	6%
Carrot	30	0	0 / 0	60 / 3	250 / 7	7 / 2	2 / 8	5g	1g	110%	10%	2%	2%
Cauliflower	25	0	0 / 0	30 / 1	270 / 8	5 / 2	2 / 8	2g	2g	0%	100%	2%	2%
Celery	15	0	0 / 0	115 / 5	260 / 7	4 / 1	2 / 8	2g	0g	10%	15%	4%	2%
Cucumber	10	0	0 / 0	0 / 0	140 / 4	2 / 1	1 / 4	1g	1g	4%	10%	2%	2%
Green (Snap) Beans	20	0	0 / 0	0 / 0	200 / 6	5 / 2	3 / 12	2g	1g	4%	10%	4%	2%
Green Cabbage	25	0	0 / 0	20 / 1	190 / 5	5 / 2	2 / 8	3g	1g	0%	70%	4%	2%
Green Onion	10	0	0 / 0	10 / 0	70 / 2	2 / 1	1 / 4	1g	0g	2%	8%	2%	2%
Iceberg Lettuce	10	0	0 / 0	10 / 0	125 / 4	2 / 1	1 / 4	2g	1g	6%	6%	2%	2%
Leaf Lettuce	15	0	0 / 0	35 / 1	170 / 5	2 / 1	1 / 4	1g	1g	130%	6%	2%	4%
Mushrooms	20	0	0 / 0	15 / 0	300 / 9	3 / 1	1 / 4	0g	3g	0%	2%	0%	2%
Onion	45	0	0 / 0	5 / 0	190 / 5	11 / 4	3 / 12	9g	1g	0%	20%	4%	4%
Potato	110	0	0 / 0	0 / 0	620 / 18	26 / 9	2 / 8	1g	3g	0%	45%	2%	6%
Radishes	10	0	0 / 0	55 / 2	190 / 5	3 / 1	1 / 4	2g	0g	0%	30%	2%	2%
Summer Squash	20	0	0 / 0	0 / 0	260 / 7	4 / 1	2 / 8	2g	1g	6%	30%	2%	2%
Sweet Corn	90	20	2.5 / 4	0 / 0	250 / 7	18 / 6	2 / 8	5g	4g	2%	10%	0%	2%
Sweet Potato	100	0	0 / 0	70 / 3	440 / 13	23 / 8	4 / 16	7g	2g	120%	30%	4%	4%
Tomato	25	0	0 / 0	20 / 1	340 / 10	5 / 2	1 / 4	3g	1g	20%	40%	2%	4%

U.S. Food and Drug Administration
(January 1, 2008)

Fig 11-1.Nutrition facts of vegetables. Source US FDA 2008.

It is suggested that eating a diet rich in fruits and vegetables that contain lutein may also decrease your risk of cardiovascular disease.

Growing food in restricted environments

11.2 Some vegetable to grow under VA

Some vegetables and the mineral and nutritious content, can be analyzed. Taste is not that important since chefs now know how improve it by and massaging them with new cooking elements like oils, salt, pepper, etc. The following vegetables are cited here because besides their nutritious feature, they can be grown using all the vertical agriculture techniques

- **Kale**

Fig 11-2 Kale

Raw Kale has a bitter taste, but the contents of vitamins A and C, minerals, fiber, antioxidants and various bioactive compounds, protein and only 50 calories make it very important.

A 100 gram portion of Kale contains:

- 200% of the RDA for Vitamin C.
- 300% of the RDA for Vitamin A (from beta-carotene).
- 1000% of the RDA for Vitamin K1.
- Large amounts of Vitamin B6, Potassium, Calcium, Magnesium, Copper and Manganese.

- **Spinach**

Fig 11-3 Spinach

with vitamin A, lutein, vitamin C, vitamin E, vitamin K, magnesium, manganese, folate, betaine, iron, vitamin B2, calcium, potassium, vitamin B6, folic acid, copper, protein, phosphorus, zinc, niacin, selenium and omega-3 fatty acids.

- **Collard greens**

Fig 11-4 Collard Greens

Loaded with Vitamin K, lutein, Vitamin C, potassium, and fiber, only yields 20 calories.

- **Swiss chard**

Fig 11-2 Kale
Fig 11-5 Swiss Chard

All the nutrient bases, including Vitamin K, lutein, Vitamin C, potassium, and fiber. It is also rich in minerals and protein.

- **Turnip greens**

Fig 11-6 Collard Greens

One serving to only 20 calories. Rich vitamin K, lutein, vitamin C, and fiber.

- **Pumpkin**

Fig 11-7 Pumpkin

Excellent Vitamin K, Vitamin C, potassium, and fiber content..

Growing food in restricted environments

- **Mustard greens**

Fig 11-8 Mustard Greens

Very low calorie content (10), high contents of Vitamin K, lutein, Vitamin C, and fiber.

- **Sweet potato**

Fig 11-9 Sweet Potato

It is very nutritious because of its big amounts of Vitamin C. It is a carbohydrate with sweet taste. It also has contents of potassium, fiber, vitamins, and magnesium.

- **Radicchio**

Fig 11-10 Radicchio

Radicchio or Italian Chicory has excellent contents Vitamin K, Lutein, Vitamin C and potassium. Only 20 calories.

- **Carrots**

Fig 11-11 Carrots

Carrots are very recommended because of the Lutein and its relation to the vision. Only 30 calories per serving. Big contents of Fiber, Vitamin K.

- **Asparagus**

Fig 11-12 Asparragus

Is a complete source of Iron, Magnesium and Zinc. Powerful nutritive features because of its contents of Fiber, Protein, Vitamin A, Vitamin C, Vitamin E, Vitamin K, Thiamin, Riboflavin, Niacin, Folate, Phosphorus, Potassium, Copper, Manganese and Selenium.

Other not so known vegetables are being object of study because they could be an interesting and nutritious alternative in future times. According to FAO, quinoa shows a promising future because of its contribution to nutrition. This organization declared year 2013 as "The International Year of the Quinoa."

Growing food in restricted environments

Fig 11-13 Quinoa

Quinoa is quickly becoming a substitute for other grains because of its nutritious properties, the bioactive contents rich in flavonoids which are very powerful anti-oxidants, higher fiber contents than other grains, free of gluten, high contents of protein and aminoacids, low glycemic index which is a measure of how quickly intakes can raise blood sugar levels, and many other healthy features.

Quinoa has similar contents of macro nutrients as other common vegetables, as shown in the table below

Table 1: Macro-nutrient contents of quinoa and selected foods, per 100 grams dry weight					
	Quinoa	Bean	Maize	Rice	Wheat
Energy (Kcal/100g)	399	367	408	372	392
Protein (g/100g)	16.5	28.0	10.2	7.6	14.3
Fat (g/100g)	6.3	1.1	4.7	2.2	2.3
Total Carbohydrate (g/100g)	69.0	61.2	81.1	80.4	78.4

Source: Koziol (1992)

Fig 11-14 Macronutrients content of Quinoa. Source Koziol 1992 via FAO

When compared to other foods, the mineral content is much better.

Table 3: Mineral content of quinoa and selected foods, mg/100g dry weight				
	Quinoa	Maize	Rice	Wheat
Calcium	148.7	17.1	6.9	50.3
Iron	13.2	2.1	0.7	3.8
Magnesium	249.6	137.1	73.5	169.4
Phosphorus	383.7	292.6	137.8	467.7
Potassium	926.7	377.1	118.3	578.3
Zinc	4.4	2.9	0.6	4.7

Source: Koziol (1992)

Fig 11-15 Mineral content of Quinoa. Source Koziol 1992 via FAO

In general, there are other possible nutritious foods coming from fruits. They can be part of any vertical agriculture process, but several other factors like weight, light and irrigation make the work complex of designing the vertical growth system.

In future, animals are going to need nutritious grass to be raised healthy and fodder, more than hay, seems to be the best option. In addition to the caloric needs it can supply, fodder is an excellent enhancer to get lower costs while obtaining a better raw nutrition.

Resources:

Five reasons you should buy Organic.
http://www.sanantonio.gov/DotGov/Archives/TabId/905/ArtMID/2753/ArticleID/1942/5-Reasons-Why-You-Should-Buy-Organic.aspx

Agricultural Health Study.
http://www.cancer.gov/about-cancer/causes-prevention/risk/ahs-fact-sheet

Pesticide–induced diseases database.

Growing food in restricted environments

http://www.beyondpesticides.org/resources/pesticide-induced-diseases-database/cancer

GMO Crops Mean More Herbicide, Not Less
http://www.forbes.com/sites/bethhoffman/2013/07/02/gmo-crops-mean-more-herbicide-not-less/

EPA. Recognition and management of pesticide poisoning
http://www2.epa.gov/sites/production/files/2015-01/documents/rmpp_6thed_final_lowresopt.pdf

USDA National Nutrient Database
http://ndb.nal.usda.gov/

Quinoa
http://www.fao.org/quinoa-2013/what-is-quinoa/nutritional-value/en/

Koziol, M. (1992) Chemical composition and nutritional evaluation of quinoa (Chenopodium quinoa Willd.). Journal of Food Composition and Analysis. 5, 35-68.

Chapter 12

Successful crops.
What is needed?

*" Don't judge each day by the harvest you reap but by
the seed that you plant"*
Robert Louis Stevenson

Depending on the side you have chosen to work (Ethno or Techno), a VA project will require the knowledge and experience achieved with traditional agriculture and the contribution of several other fields, where the interdisciplinary concepts will play an important role.

The main goal of this book is to show several possibilities to grow food, but since the vegetable varieties and interests can be uncountable, it is required to develop the applicable design of growing system for each case. The importance of the experience relies in sharing any acquired knowledge to allow others to have a smooth and easy replication.Something is sure, every successful crop must be protected beforehand and the first risk condition to be avoided is that depending on the location and the climate, the crop must have either a roof on top or be completely indoors.

Architecture and building designs will be necessary beyond the focus on comfort or bioclimatic conditions. The new places must allocate several technologies which will help in future on any plan growth.Climate change will bring heavy rains and droughts so harvesting, treatment and re-usage of water will be mandatory.

Technologies for irrigation will be required. Not just bringing water but controlling the specific required amounts and the application when the plant will assimilate the most. Since the restriction is to guarantee the best usage, the aeroponics could be a better approach. Under this approach there is a good chance to develop growing systems with manual irrigation so some basic understanding on the water requirements for every planted vegetable will be required. Some of the designs show here, have proven to be very efficient in saving water and only using the minimum, but there is much research to be done first.

Efficient use of natural light will require a different technology which must be incorporated, not only in the building design, but on the light spectrum by filtering only the portion which is beneficial to the plants. Illumination is a very important factor for the photosynthetic process of plants so there will be a need for natural and artificial light. There are a lot of discussions regarding the use of artificial light and its efficiency in growing plants indoors, but this is covered in later chapters.

Most of current agriculture requires of chemicals, in the form of pesticides, herbicides and other products to protect the plant against chemical and biological enemies and species. The success of the crop will depend on the control of the environment so confined spaces will be a must condition. With the future limitations, humans and vegetables will fight for space so special conditioning will be required because of the possible negative effects on each other. The only possible interactions will happen probably in the initial and final stages of the plant growth.

All the processes will require energy from a power source so there are concerns regarding the type of energy to use and the possibility for having it "clean" or environmentally friendly.

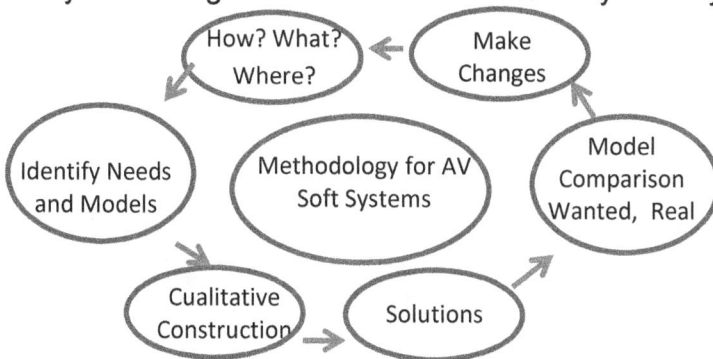

Fig 12.1 Soft Systems Methodology for Vertical Agriculture
Growing food in restricted environments

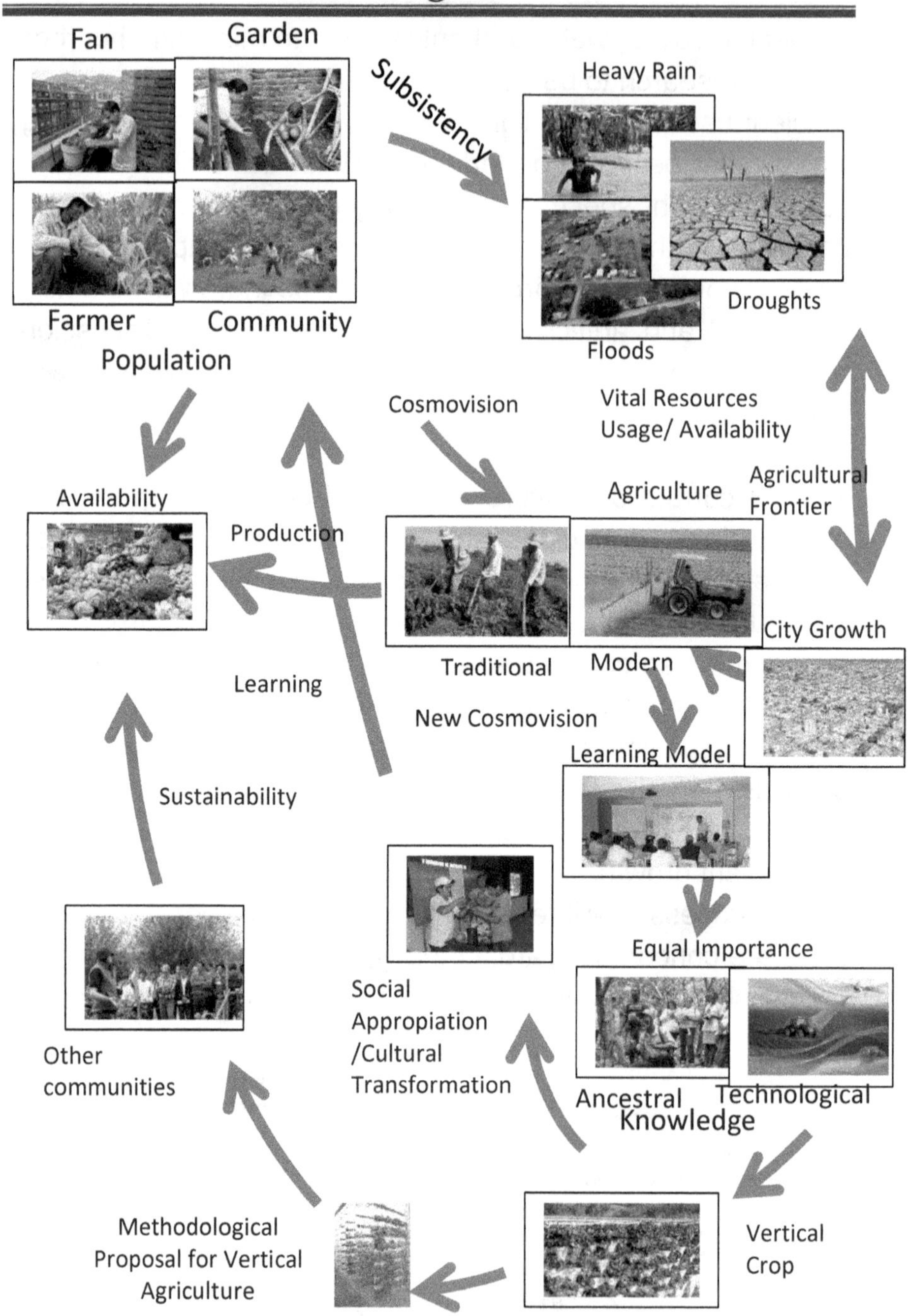

Vertical Agriculture

Fan   Garden

Farmer   Community
Population

Subsistency

Heavy Rain

Droughts

Floods

Cosmovision

Vital Resources
Usage/ Availability

Availability

Production

Agriculture   Agricultural
Frontier

City Growth

Learning

Traditional   Modern

New Cosmovision

Sustainability

Learning Model

Other
communities

Social
Appropiation
/Cultural
Transformation

Equal Importance

Ancestral   Technological
Knowledge

Methodological
Proposal for Vertical
Agriculture

Vertical
Crop

Fig. 12-2 Enriched picture of VA. Understanding the approach.

Growing food in restricted environments

Agriculture and women?

The study will be directed mostly to women to understand their needs, the kind of situation they have and the way a vertical grow unit can affect her future life. It´s important to know how much time they have or spend daily on farming labors

Resources:

Checkland, Peter. Soft systems methodology

Will Venters, Tony Cornford and Mike Cushman
Department of Information Systems, London School
of Economics and Political Sciences, London, UK
Knowledge about Sustainability: SSM as a Method for
Conceptualising the UK Construction Industry's Knowledge
Environment
http://personal.lse.ac.uk/cushman/papers/KnowledgeAboutSustaina
bility.pdf

Indigenous Knowledge,People and Sustainable Practice.Douglas
Nakashima and Marie Rou´e UNESCO,Paris,France
Centre National de la Recherche Scientifique, Paris,France
http://portal.unesco.org/science/es/files/3519/10849731741IK_Peopl
e/IK_People

Chapter 13

Water and Irrigation for Vertical Agriculture

"Water is the driving force of all nature"
Leonardo Da Vinci

Vertical Agriculture

Any crop will require water and preferably with nutrients. With the exception of precision agriculture the water is administrated either manually or through pumping systems, but none of them have any major control on the efficiency of the process.

When the plants are on soil, a lot of water and nutrients are wasted because the solution filters to deeper ground, sunlight could produce evaporation, heavy rains will move the soil and diminish the concentration of nutrients.

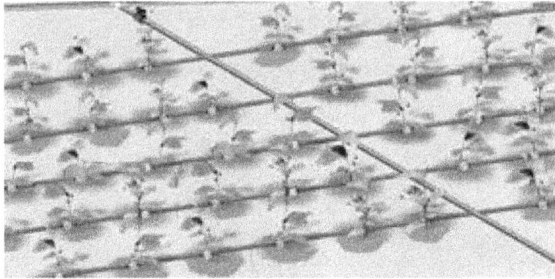

Fig 13-1Classical irrigation of plants

It´s quite important to know how to bring water to the plants mainly if the system is implemented with VAGUs Techno, because manual irrigation is always a subjective concept and can´t be controlled precisely.

13.1. Manual irrigation

These systems require a person to deliver the water or nutrient solution to a plant. They are high labor inputs for highly extended crops. Manual irrigation has evolved to small-scale drip irrigation, which uses buckets with manual valves to source small flows or drips to the roots.

13.2. Automated Irrigation

Growing food in restricted environments

Vertical Agriculture

A basic irrigation system includes the following elements as shown below:

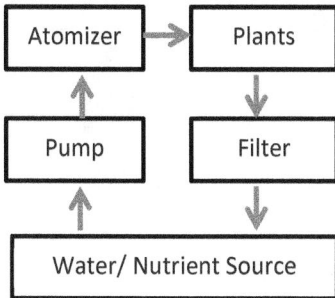

Fig 13-2. System 1

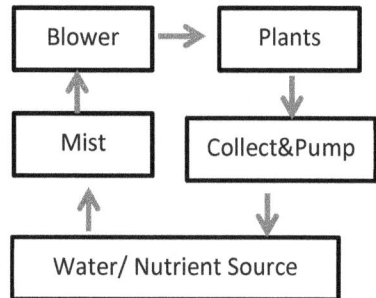

Fig 13-3. System 2

Both systems have different efficiency and application. The process starts by preparing solution with water which has diluted some required set of nutrients. The composition for the nutrient solution is defined by kind of vegetable selected for the crop and the need for successful results. This knowledge is not part of this book and it´s let to the appropriate disciplines. In both systems, the container for the water/nutrient source must be fully sealed to avoid any losses by temperature. Some aeration might be required.

13.2.1 System 1

This system can be considered as a hydroponic system, because of the use of fluids. It the nutrient is fully diluted the irrigation is exactly as it were water.

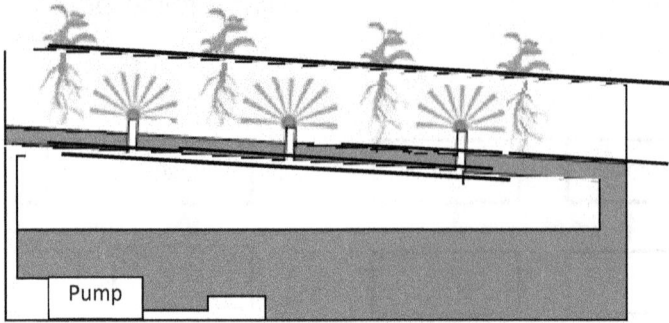

Fig 13-4 Hydroponic system

The pump is an electro-mechanical device to impulse water from one location to another. It's main purpose is the use of pressure in order to be able to transport the water. In the system on the left, with a Vertical Agriculture production, the main objective is to transport the nutrient solution form a container located at a lower level up to an atomizer which sprays the plants roots.

Fig 13-5 Irrigation pumps

Two parameters are very important and decisive when selecting the pump:

- The number of liters per minute (lpm) or gallons per minute (gpm) it can pump out.
- The maximum head. This means the maximum height it will bring the water up with the same pressure.

Growing food in restricted environments

The atomizer spreads the solution in several directions allowing the contact with the roots, which are the best way to feed the plant. The part of the solution that is not absorbed (nearly 95%) falls to a filtering system and can be used again.

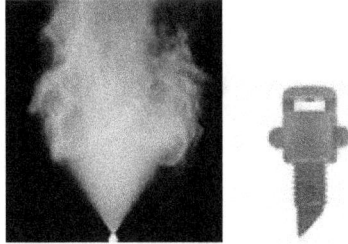

Fig 13-6 Atomizers

13.2.2. System 2

Under some research conducted by NASA (nautical Aerospace Administration) form USA, some plants' roots were irrigated by a mist instead of a liquid medium as shown in system 2. The results were far more efficient because the roots have better assimilation of the nutrient droplets.

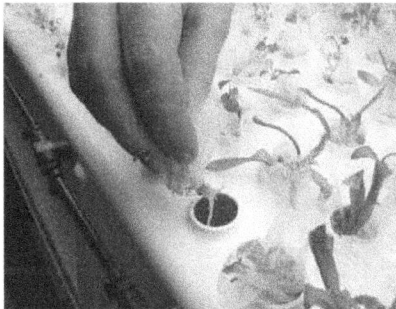

Fig 13-7 Aeroponic System. Source NASA

Growing food in restricted environments

These plants have developed healthy root systems without soil in a rapid-growth aeroponic system. (Image credit: AgriHouse, Inc.)

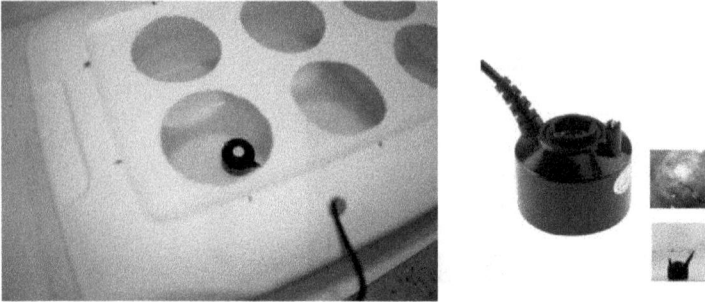

Fig 13-8 Aeroponic system

This new technology got then name of Aeroponics. It has the advantage of using much less water/nutrient solution so the energy and the efficiency is enhanced. A fogger is a device used to sublimate the nutrient solution into a fine mist which can be blown (transported) directly to the roots of the plants.

Fig 13-9 Roots growing with mist

The electrical requirements are less than using pumps and there is almost no waste of resources compared to hydroponics. Aeroponic systems are very suitable for food production in space travels.

Control elements for both systems are about the same hence they have almost the same cost.

13.3 Nutrient Solution

Both systems will require a nutrient solution. Research centers can use or develop these solutions for specific needs or interests. Farmers and fans of agriculture can use commercial nutrient solutions which are sold for general agriculture applications for home or commercial systems. They are loaded with the essential nutrient molecules in order to help a general plant to grow healthy.

It is necessary to be careful, not all the nutrient solutions will be appropriated for some plants. The solutions can have different effects on taste, growth factors, color, leaves formation, etc.

There are some new techniques, among them the Microbiomes, which are important because of the future they represent for vertical agriculture and plant health. A Microbiome is micro and complex community of microbes.

Microbes are present in any environment. Some of them, like bacteria or fungi, can be beneficial when added to plans or growing media. There are several methods either to increase the amount of microbes or to add a totally new population: by mixing water with microbial agents and spraying them on roots, soil plants, by bringing the plant to climate chamber where the conditions help the microbial growth or mixed with other growing elements.

Fig 13-10. A grower clips the leaves of plants grown in the openings of an aeroponic growing chamber. (Image credit: AgriHouse, Inc.)

Fig 13-11 Hydrogel. A film to grow food. Professor Yuichi Mori, Waseda University , Japan 1995

Growing food in restricted environments

Vertical Agriculture

Professor Yuichi Mori, Waseda University has developed a polymeric membrane technology to grow vegetables.
In 1995, he launched Mebiol Inc., the first and only company to commercialize a medical membrane–based plant cultivation technology, Imec. The picture shows the amazing advantages of this new technology.

Resources:

Water saving.
10 ways farmers are saving water.
http://www.cuesa.org/article/10-ways-farmers-are-saving-water

Irrigation.
https://afsic.nal.usda.gov/soil-and-water-management/water-saving-irrigation-techniques

Microbiomes.
http://learn.genetics.utah.edu/content/microbiome/

Aeroponics.and NASA.
http://www.nasa.gov/vision/earth/technologies/aeroponic_plants.html

Aeroponics.
http://www.agrihouse.com/

Film Farming.
http://www.mebiol.co.jp/en/

Growing food in restricted environments

Chapter 14

The new architecture

"The decline in arable land, ongoing global climate change, water shortages and continued population growth could change our view of traditional farming from soil-based operations to highly efficient greenhouses or urban farms." – ALLEN WASHATKO, TKWA

The new architecture should strive for some main focus:

Green construction, fully integrated with the environment, with Water Harvesting and at least, it should include a sustainability wall

The focus must be directed to houses and buildings.

The green construction refers to the way new constructions should be made, under specific ecological tasks which take care of any ecological consideration.

Kenneth Yeang, a very renowned architect with more than 70 international awards since 1975, proposes a different approach for the integration of agriculture and population:

"Instead of hermetically sealed mass produced agriculture, plant life should be cultivated within open air, mixed-use skyscrapers for climate control and consumption,"

Yeang is one of the promoters of bioclimatic design for buildings. The construction and occupation are somehow built using sustainable and carbon-neutral practices. In addition to the possibility of using specific features on the building structure in order to help with the interior temperature, he tries to include agriculture as part of the strategy for climate control.

14.1 Ecologic urbanism.

Urbanism is a consequence of population growth. Some professional are changing the classic architectural designs and turning them into a more green construction which

Growing food in restricted environments

combines ecology, environment, climate, design and urbanism.

One of the new proposals is from the architect Mitchell Joachim, who in addition to his private firm, teaches at Columbia University and Parsons.

Among several other interesting proposals he is proposing a new architectonic design which can combine both architecture and biology. Just imagine the possibility to literally grow a house or a building by planting trees together. This tree shaping (or tree training), can be used to build useful structures.

Fig 14-1. Pleaching trees

The branches will interweave together and in some cases, when there is a close contact, they will grow together. The process is called "pleaching".

Growing food in restricted environments

FIG 14-2 Fab Tree Hab. Mitchell Joachim, Lara Greden   and Javier Arbona

The total structure can be the basis of a house which is completely green.

At his company they have developed a lab where they grow extracellular matrix from pigs to be used as the material to build, via simple 3D printing, difficult shape objects (shoes, leather, handbags, etc), without affecting living creatures. This is a very important remark from their job.

These techniques can be applied to vertical agriculture to grow green walls which are really green, and having a wall that could be food at the same time is an alternative worth to work on.

14.2. Urban Farms

14. 2.1 Urbanana

Urbanana is an urban farm of vertical agriculture where the main crop is banana which is located on the façade. The Footprint of this building is 1.290 m2 and 24 Meters tall and six floors, with an available cultivation surface of 3, 8Ha. The

Vertical Agriculture

remaining volume is used for a restaurant, a research labs, a space for expositions and a boutique.

Located in Paris, France, it was made by SOA architects.

Urbanana at night

Bananas behind the glass

Fig 14-3 Project Urbanana.

Growing food in restricted environments

25

Fig 14-5 The whole building design

The project can be considered a vertical agriculture farm, and it's more than simply a greenhouse with transparent glasses. It is a mechanical structure, with technical features to allow full commercial exploitation. The light can be either natural or artificial and with help of artificial heating, the banana trees have normal production throughout the year.

The implementation costs are very high, making them prohibitive for any commercial application, but this is a good beginning to explore more alternatives of food production. It is quite possible that when more systems like this one are built and the technology becomes available, prices will go down.

14. 2.2 Pasona urban farm.

Located in Tokyo and made by SOA and Kosono Designs this urban farm was made by renovating an old building.

Fig 14-6. Project Pasona

To ensure sustainable future food production this building focuses on providing education and training on traditional and urban farming to students and small farmers, since the activity of farming has been notoriously decreasing in Japan.

The building counts a modern farming technology to maximize crop yields. : lighting (metal halide, HEFL, fluorescent and LED lamps), automatic irrigation system and an intelligent climate control system (for temperature and moisture) to guarantee yearlong crops.

In harvesting and maintenance of crops there are groups of specialists and employees from the building. By participating in these processes people get engaged in sustainable practices which end with the consumption of their own products right there in the building's cafeterias.

14. 3 Wall of Sustainability

Not all of us have a garden or patio but we all have a wall. The idea is quite simple, every place and mainly the ones about to be constructed should include a wall specifically designed or conditioned for food produce.

Vertical Agriculture

What is this concept? It´s a wall with all the necessary to grow vegetables: sunlight exposure if possible or Ac outlets for connecting devices, irrigation and water harvesting, soil or soilless holding systems, etc.

At some point, under the climate change effects, if would be nice to have a simple and friendly setup at home ready to grow vegetables.

Modern buildings are designed with tight spaces in mind. Designers sacrifice green zones for living spaces, because it is the last one which is more profitable.

If sustainability wall were part of a policy, especially for builders of low income community, a new era of nutrition and agriculture could arise. If people have fewer efforts in taking care of their agriculture, it is possible that they were being more attracted to practice it. This condition is applicable to urban or rural habitants, with basic or no knowledge of agriculture.

The walls are Vertical so the sustainability walls are considered as part of Vertical Agriculture. Conceiving the wall in a home from the stage of design will be cheaper, better and faster to implement.

Fig 14-7. Wall adaptation to grow food.
Source: thelivinggreens.com.
Edible green walls 2015

The future climate scenarios, especially in developing countries, will force population to grow food in any possible space they have. It would be wonderful if architects and civil engineers could include all type of sustainability walls in their design and construction projects.

14.4 The home/farm of the future with Vertical Agriculture

Although some aspects will be covered in the next chapter, it is important to mention how the new architecture should get involved in water harvesting, lighting and agriculture.

Water is going to be scarce in future time and potable water is going to be something everybody will need so, why waiting to include solutions that can save lives in future? If all the new designs could have all the elements to harvest, let´s say, rain water, by bringing it where potable water is used for purposes other than human consumption, the world would be a much better place where sustainability plays a key role. There is much to do, but unfortunately architects and hydraulic designers of houses and building don´t pay much attention to the requirements for water harvesting and the task becomes very expensive and complex at later stages.

Growing food indoors has several associated costs which can be very high. Saving water and making a rational use of it should be the new policies for every construction, looking to increase the technology for in-situ close loop systems, which are far more efficient than the normal systems we use today.

Day lighting is a real need on today's architecture. It can be really beneficial for humans and plants, as proposed with this new approach of Vertical Agriculture.

Imagine a world in 50 or 100 years from now. We had learned the lessons from our mistakes and as result, we had conceived innovative solutions. The following is a suggestion of the future home conceived to be a place to live and develop food.

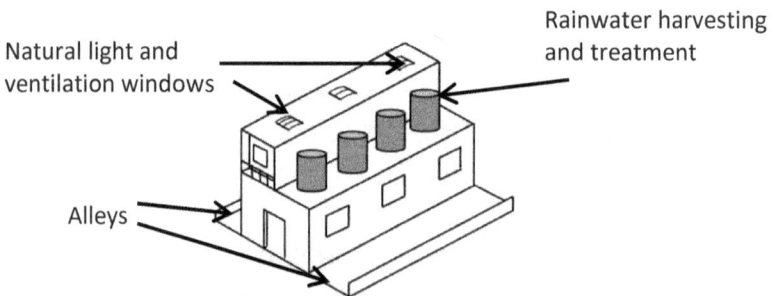

Fig 14-8 House of future

Growing food in restricted environments

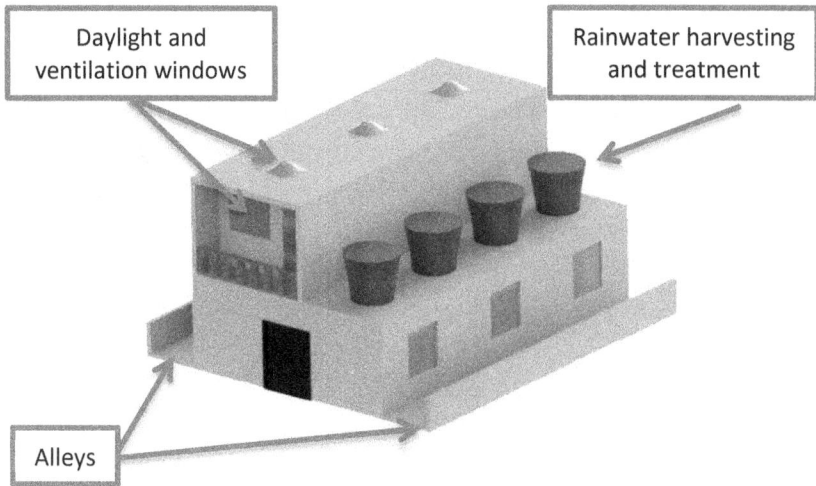

Fig 14-8 House of future

Now it´s time to develop a structure to support the home and any possible agriculture. For that purpose, the way bearings are designed to operate in mobile systems will help to understand the operation of the proposed structure. The next figure show the elements of a ball bearing

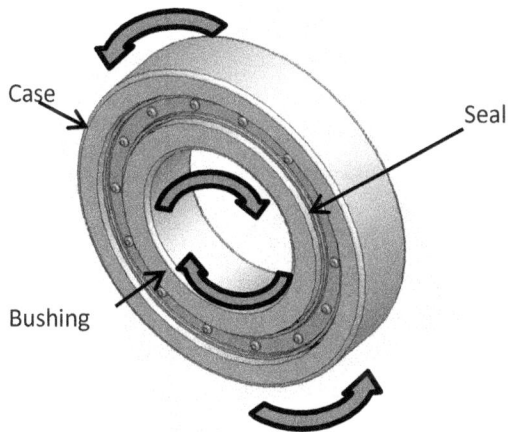

Fig 14-9 Working Principle of bearings

Growing food in restricted environments

The external (Case) or internal (Bushing) cylinder can rotate freely, so they can be attached to any mechanical system. In our case, we need the external part to rotate so; the bushing is attached to a solid fixed structure. On the case we add some small cylinders, as shown on the next figure, which connect the exterior with the space between the case and the

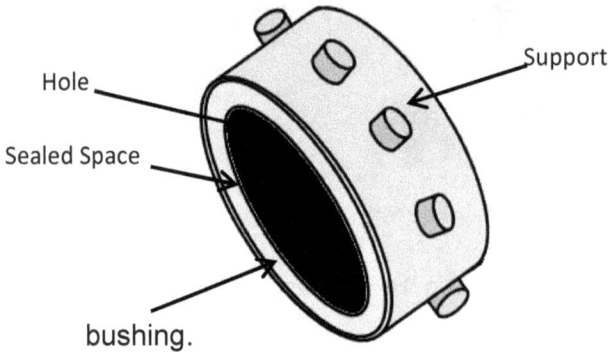

Fig 14-10 Home/Growth module

If needed, it is possible to add more modules like this one, in order to have longer capabilities as shown below. The materials used to build these modules are not important for now.

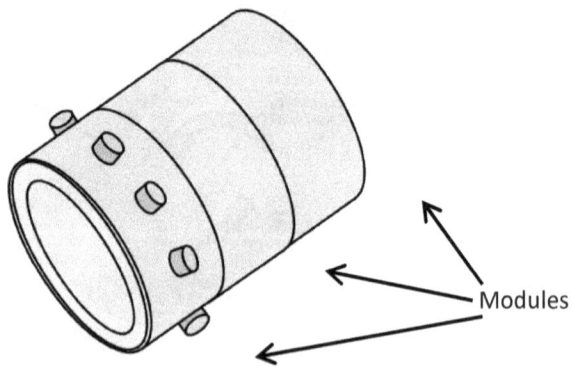

Fig 14-11 Home/Growth module assembly

Growing food in restricted environments

Vertical Agriculture

This structure has to be as big as need to allow having a comfortable home inside, just like the one we have defined earlier. It also has some small drains for rain water harvesting and some spaces to allow sunlight to pass thru, in order to allow using natural light on the interior.

The house inside the structure would look as shown in the next graphic.

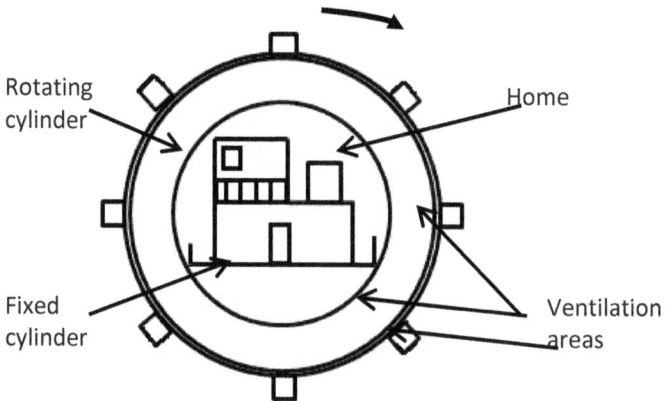

Fig 14-12 Home/Growth module. Front view.

The home should fit inside the inner cylinder so there is here an interesting combination of space use. This fact will give an idea of how big this system is going to be. The alleys at every side of the home will allow some mobility from module to module. The extra space helps with the ventilation and any possible heat dissipation of the interior for a bioclimatic design and more comfortable living conditions. The thickness of the seal or space between the two cylinders will provide an extra and very good thermal isolation.

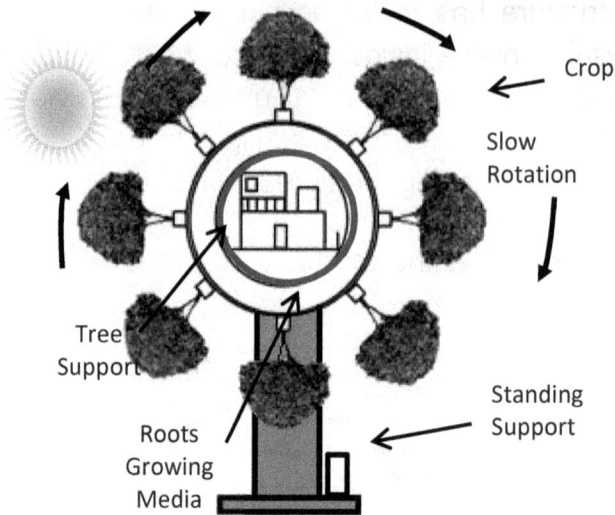

Fig 14-13 Home/Growth module. Front view

Next step is to add an external crop, as shown in the above picture. Complete agriculture can be developed on the surface of the structure. Trees (food) as big as desired can grow under the limits of height of the system. The crop will slowly rotate to allow even amount of sun exposure of plans or trees. This constant (or programmed motion) won't have any shaking effect on the trees because it's almost imperceptible. The trees also act as a roof that provides shelter and shadow for smaller vegetable species.

The plants always try to reach the sun. This phenomenon is known as phototropism is expected but probably not noticeable at all.

The roots will grow inside a chamber which offers several advantages: no evaporation, no land degradation, not hydric stress on rivers, minimum filtration, any applicable soil or soilless technology, easy access to roots to check correct

Growing food in restricted environments

nutrient assimilation, study or prevention of damage or simply for fruit collection (like in the case of potatoes).

The harvesting is also easier since people at ground level can check inaccessible tree branches.

Moving the external cylinder only requires an initial torque because the rest becomes an inertial motion system which requires much less energy. This is useful in sizing a motor which in turn could have solar panel as main energy source.

Rain water can be collected at several levels: external cylinder, at the home and at the ground. Since this system saves water, it is quite possible that a heavy rain of just one day is enough for a complete food production cycle of the vegetable species.

Regarding the occupied size, just as an example, the following analysis can be performed:

 A land of 10.000 square meters (100mx100m) will be used as an example. It can have up to 400 apple trees of 6 meters tall and occupying a square of 5 meters of side each.

The Vertical agriculture equivalent land can be calculated as follows:

The perimeter or length of the circle is 100m =2xΠxR. Calculating for R, a value of R= 15,9 is obtained which can be rounded to 16 meters.

The height of two trees must be added to calculate the maximum horizontal distance:

Vertical Agriculture

Total Horizontal Distance= 2x 6m (height) + 32 m (Diameter) = 44m.

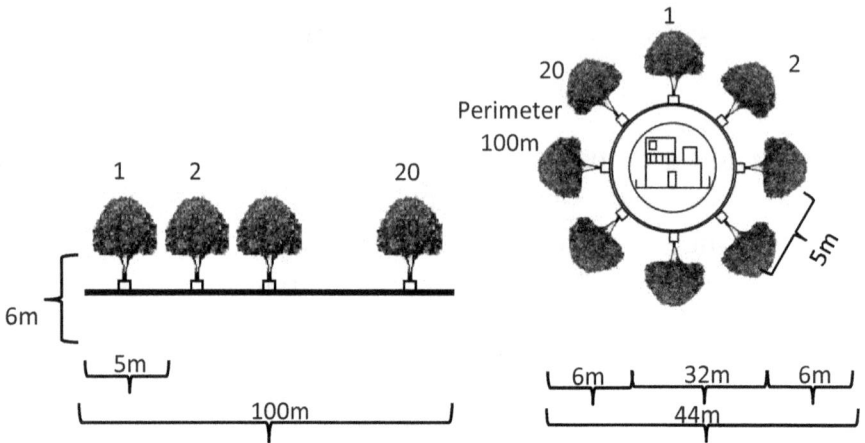

Fig 14-14 Home/Growth module. Same production much less space required.

The new total distance required to have a Vertical Agriculture Crop for the same horizontal space, with the same tree is around 44 meters which represents 56% of savings compared to the land required for the horizontal distribution.

The graphic shown below shows something very important. Under the same parameters of distribution of trees (separation of 5 meters and 20 trees per line), by increasing the diameter of the cylinder in just 2 meters (or increasing the radio in one meter from 16 to 17 meters) the number of trees goes from 400 to 427.

In other words, there is a linear correspondence with the number of trees. By increasing the radio in a 6%, the number of trees increments the same amount.

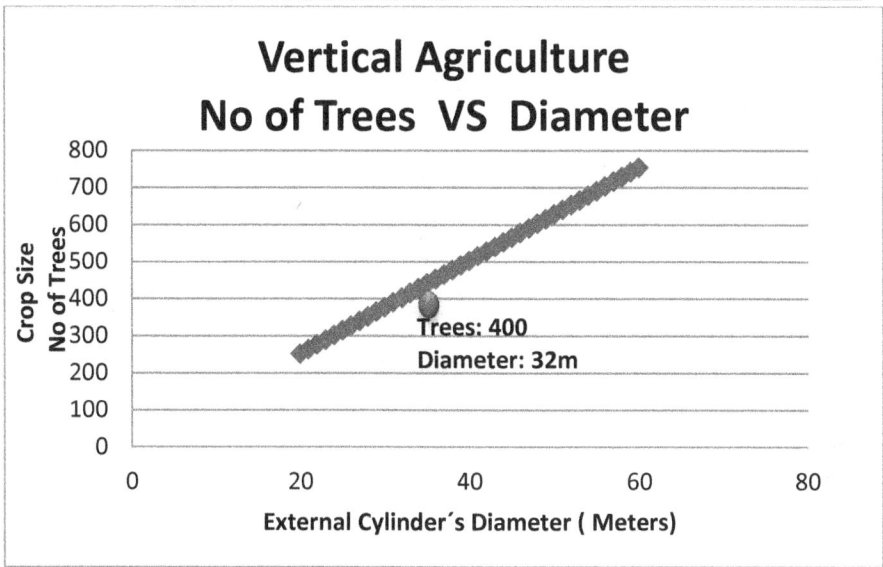

Vertical Agriculture
No of Trees VS Diameter

Crop Size / No of Trees — External Cylinder's Diameter (Meters)

Trees: 400
Diameter: 32m

Fig 14-15 Home/Growth module. Production
according to module´s diameter

A very important remark: by a simple change to use a Vertical Agriculture, it is possible produce almost DOUBLE of what is being produced with Horizontal Agriculture, in addition of the smarter use of vital resources. With smaller height vegetable species, it is possible to have even more production.

Some other considerations are also possible. For instance, if the maximum height of the crop is decreased or the distance between the plants, by selecting a different vegetable (for example Sugar Cane) the total horizontal space is less so, there is a huge increase in any crop production by using Vertical Agriculture.

Proving that the above assumptions are right is just part of the future, please wait for more or simply join us!!!

Growing food in restricted environments

Resources:

Pleaching trees
http://www.instahedge.com/product/pleached-trees/
http://www.telegraph.co.uk/gardening/plants/trees/8916079/Pleaching-The-art-of-training-trees.html

Fab Tree Hab
http://www.archinode.com/fab-tree-hab.html
http://www.archinode.com/Arch9fab.html

Urbanana
http://www.soa-architectes.fr/en/

Pasona Urban Farms
http://konodesigns.com/portfolio/Urban-Farm/

Edible Green Walls
http://thelivinggreens.com/
http://growediblewalls.com/

Chapter 15

Water Harvesting

"We forget that the water cycle and the life cycle are one".

Jacques Cousteau

Climate Change will bring heavy rain periods so, water is going to be so abundant in some places that one of the ways for adaptation is learning to look at it as an advantage rather than just a problem.

Rainwater harvesting had been overlooked, but nowadays it is becoming very popular because there is an impending need of using water resources more wisely.

Collecting and using rainwater will bring several advantages:

- Less hydric stress over rivers.
- Addressing of potable water to population rather than simple uses like toilette flushing.
- Lower costs of water utility bills.
- Friendly and responsible environmental policy.
- Allow others to have potable water available.
- Reduce the costs of energy and chemicals to purify water.

15.1. Important considerations.

It is necessary to be very careful with this approach, because in some countries, collecting of rainwater can take you to legal problems. A permit (which always has a fee) from the local authority of building inspection or plumbing division is required if the downspout is connected to the sewer system. In addition to the intended use of the water (irrigation, flushing, etc.) you have to submit plans of the system.

Incredibly, there is a rural story about Gary Harrington of Eagle Point was sentenced to jail and the payments of very expensive fines in Jackson County (Oregon, USA) for rain water usage and collection. He applied and obtained the

Growing food in restricted environments

permits from the Water Resources Department but later, they reversed the decision based on a controversial law from 1925, which establishes that all of the water in the state of Oregon is public water so; any person has to acquire a permit before giving any use to it.

In some other places, if the downspout is disconnected from the sewer system, sometimes there is no need for permits, as long as your rain container fulfills some requirements like:

- Under a given number of gallons (for example 4,000 gal)
- Height to width ratio is less than a given proportion (For example 2-to-1) and if it has the risk for a person to enter, it has to be labeled with preventive signs or messages like "Confined Space"
- Does not require power or makeup water supply connection
- Supported directly on grade;
- Used exclusively for non-spray irrigation.
- Container with a sticker, stencil or label: "Non Potable Water". "Do Not Drink"

Fig 15-1 Water Collection at homes

15.1. Rain water systems for homes.
Growing food in restricted environments

More sophisticated systems, which collect rain water, are becoming a god option for places where droughts are common or where the home owner wants to contribute to make a responsible use of it.

If the storage tank is underground, it is necessary to pump it up to a smaller tank, located in the higher level. The height helps to supply the water by gravity, with enough pressure to work on lower levels: This water can´t be used for human consumption so it can´t be connected to the potable water system and new ducts have to be added to drive the water to specific places around the home. In the case below the rain water could be used to flush the toilettes, or other simple and not demanding applications like irrigation of gardens and clothes washing.

Storage tanks could grow bacterial contents so, in some cases, some chlorine might be required to eliminate the algae or the fish smell of contained waters.

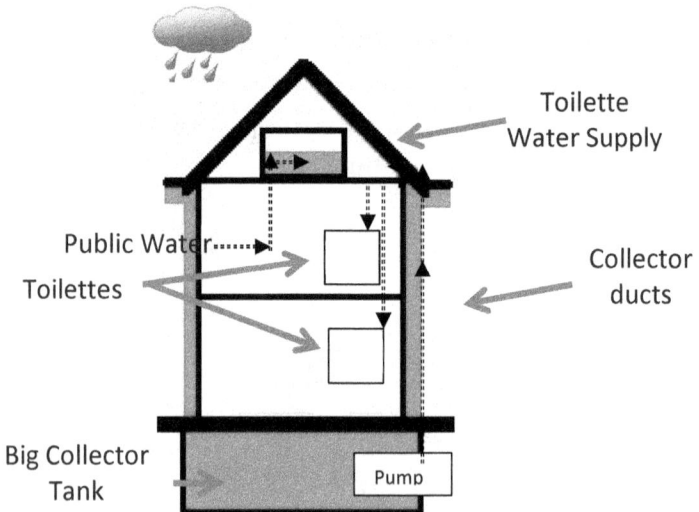

Fig 15-2 Rain water harvesting system for homes.

Growing food in restricted environments

In commercial and high populated buildings, the application could be very interesting since the savings of water could be considerable. The mechanical and control elements are about the same as used in home applications.

One of the main mistakes the new architecture is making is that they only consider rooftop but not wall rainwater collection. The walls can collect almost as much water as a roof, increasing the amount of water to be used

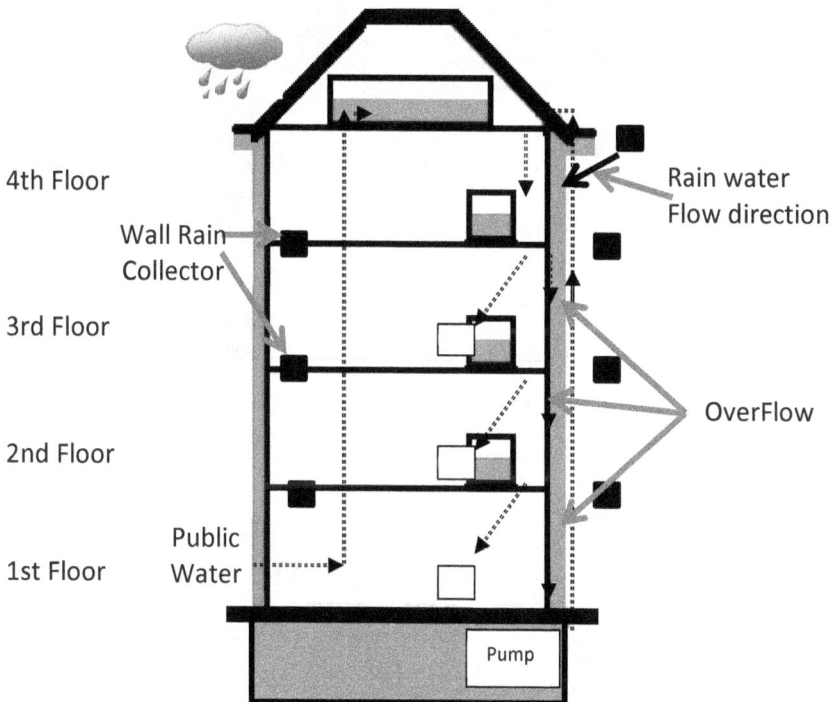

Fig 15-3 Rain water harvesting system for buildings.

Growing food in restricted environments

Another very common mistake is the use of pumping to bring water to roofs, but if the water is already up, why not to take advantage of this energy saving practice?

15.2. For urban centers and new constructions

Traditional systems use the following elements:

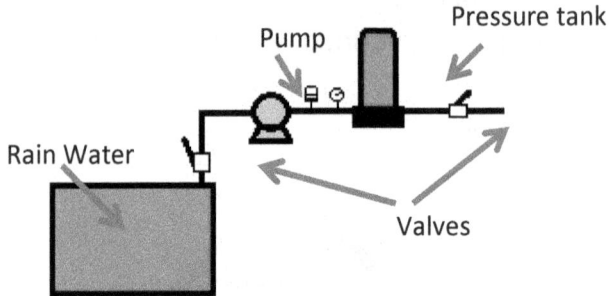

Fig. 15-4 Basic water pumping system

From the energetic point of view, although this system is operational it is also considered very inefficient. The reason is related with the fact that the pump must work continuously and with a direct electrical connection. Because of this, the starting currents of the motorized pump are very demanding and usually higher than the power supply capability. In some cases the power supply is based on solar panels which force to oversize the design. There are solutions for this problem, like the soft starters, but they are not necessarily popular because they represent an additional cost.

In the next graphic, the blue lines represent water flow and what is shadowed is what has been successfully tested and implemented.
Dotted lines are used to represent electrical connections and energy flow.

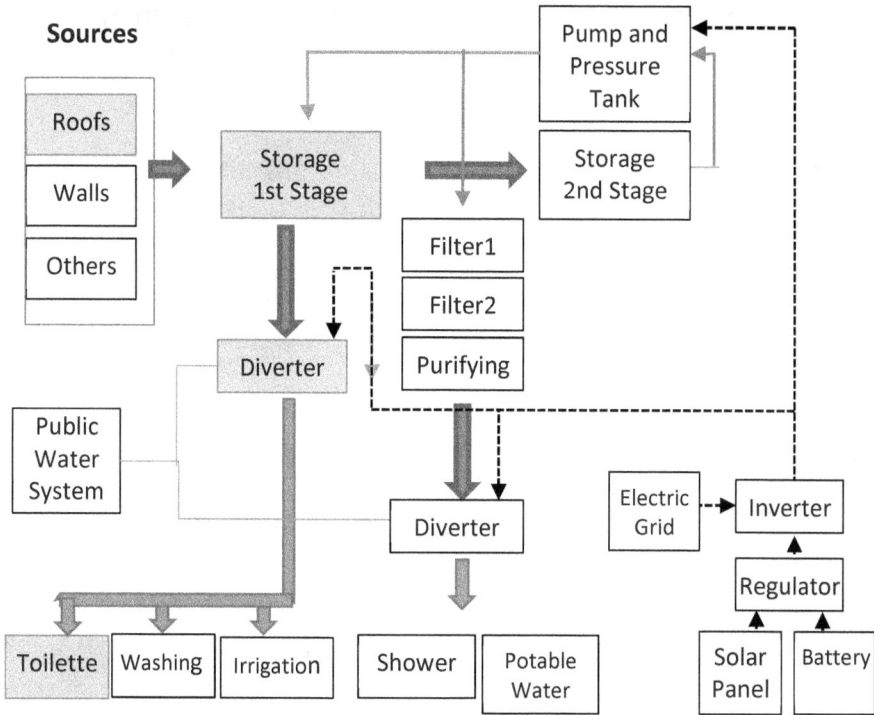

Fig 15-5 Total Design for a Rain Water Harvesting
System. Source: Orlando Charria 2013.

15.3. Frequency of rain events

Climate change comes with random periods of rains and droughts. And variable durations which range from minutes to hours While rain can be common in some countries in other countries the effects will be completely opposed.

The frequency of rain events gets importance when selecting the final control element to be used to save water. The electrovalve.

In cases of automated systems, the electrovalves have a coil which must be energized for them to work. If the mechanical

Growing food in restricted environments

part allows flow of water without energizing the coil it is called a Normally Open (NO) electrovalve. If the valve, in normal conditions with absence of voltage on the coil, blocks the flow of water it is called a Normally Closed (NC) electrovalve.

Sustainable solutions will require the minimum waste of electric energy to be applied to the coil. It is a kind of programmability which is not common in commercial units. The less need you have to energize the valve´s coil the better, but this condition is linked to the reigning weather conditions

The next table summarizes the type of valve needed for this particular application, using this same design.

More frequent	Valve Type	Normal Status
Rain	NC	De-energized
Rain	NO	Energized
Drought	NC	Energized
Drought	NO	De-energized

Fig 15-6 Rain water and required valve.

Fig 15-7 Commercial system for rain water harvesting. Source Charria 2014.
Growing food in restricted environments

Fig 15-8 Prototype of water harvesting valve
(patent in process). Source Charria 2014

There is a lot of research on this topic and a very interesting free software tool to calculate the rainwater which can be harvested was developed by the North Carolina State University. The tool is available for free downloading.

Resources:

Rain water calculations
http://www.bae.ncsu.edu/topic/waterharvesting/RainwaterHarvester_2.0.zip

Collecting Rainwater
http://www.cnsnews.com/news/article/oregon-man-sentenced-30-days-jail-collecting-rainwater-his-property

Water harvesting valve and projects
www.hi-tech.co

Growing food in restricted environments

Chapter 16

Modern energy generation

" We must move away from our dependency on fossil fuels, and I´m glad that GM has invested over $1 billion on hydrogen fuel cell cars to meet this goal"
Albert Wynn,
American politician.

16.1 Basic concepts of Fuel Cells.

Although the Fuel cell was invented in 1838, only in recent years it has been a new focus of attention, because of NASA and its space era applications.

The Fuel Cell is an electrochemical system capable of generating electricity from a combination of Hydrogen and Oxygen. The big advantage is that they produce just heat and water as sub products of the process.

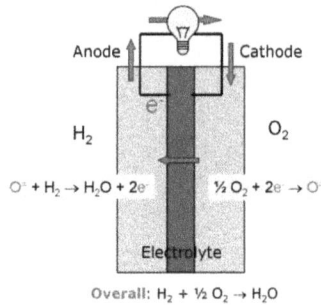

Anode   Cathode

e^-

H_2 O_2

$O^- + H_2 \rightarrow H_2O + 2e^-$ $\frac{1}{2}O_2 + 2e^- \rightarrow O^-$

Electrolyte

Overall: $H_2 + \frac{1}{2}O_2 \rightarrow H_2O$

Fig 16-1 Working principle of Fuel Cells

The fuel cell is formed by three elements: the anode, the cathode and the electrolyte. The Hydrogen is supplied as fuel to the anode. On the other side, air is supplied to the cathode. When a load (let´s say a lamp) is connected between anode and cathode some electrons will flow through it which in time it´s an electric current. In general terms the Fuel cell is producing an electric current which is turned into energy when a load is connected.

Some electrons from the hydrogen are combined with the oxygen molecules from the air producing water which can be recycled later. The current through the load also generates heat.

Growing food in restricted environments

Vertical Agriculture

Because of its electrochemical nature, there is not fuel combustion at all so; the Fuel Cells are friendly to the environment, free of noise and very energy efficient.

Hydrogen is the fuel for the Fuel cell. It´s molecule is always combined with other elements like the oxygen in water and the carbon in methanol and ethanol. There are several ways to obtain Hydrogen: steam reformation (from methanol, ethanol propane, etc.), water electrolysis ad enzymes in combination bacteria and algae.

16.2. Fuel Cell technologies.

Many commercial fuel cell types can be found: Alkaline Fuel Cell (AFC), Proton Exchange Membrane (PEM) fuel cell, Direct Methanol fuel cell (DMFC), Molten Carbonate fuel cell (MCFC), Phosphoric Acid fuel cell (PAFC) and Solid Oxide fuel cell (SOFC).

Four parameters are important to consider here: Electrolyte, catalyst, operating temperature and energy efficiency.

The table below shows all the different Fuel cell and its components

Type of FC	Electrolyte	Catalyst	Operating Temp (^{0}F)	Energy efficiency (%)
PEM	Solid polymer membrane.	Platinum	175-200	40-60
DMFC	Solid polymer membrane.	Platinum	125-250^{0}F	40
AFC	Potassium	Non-	225-475^{0}F	60-70

Growing food in restricted environments

	hydroxide solution in water.	precious metal catalysts		
PAFC	Liquid phosphoric acid ceramic in a lithium aluminum oxide matrix.	Carbon-supported platinum	350-400	36-42
MCFC	alkali (Na & K) carbonates retained in a ceramic matrix of $LiHO_2$	Lower-cost, non-platinum group	1,200	50-60
SOFC	Ceramic, yttria-stabilized zirconia (YSZ)	Lower-cost, non-platinum group	1,800	50-60

Fig 16-2 Fuell cell types. Source: http://www.fuelcells.org/

Other Fuel cell types in current research and development:

Regenerative Fuel Cells (RFCs): Hydrogen and Oxygen are obtained from water by the electrolysis process. Since the Fuel Cell produces water again; the system becomes a closed loop solution.

Zinc Air Fuel Cells (ZAFCs): Competing with lead-acid batteries, zinc pellets and air are combined to produce electricity.

Microbial Fuel Cells (MFCs): Microorganisms can change organic matter into fuel.

Growing food in restricted environments

Fig 16-3 Commercial Fuel cells. Courtesy of Horizon Fuel cells

16.3 Successful cases

A company founded in 2001 in California (US), Bloomenergy has made a commercial SOFC fuel cell which can act as an off-grid power source. Many companies are already using the product to power the computer server room because of its advantages: the cell can have a working temperature of 980°C without damages or maintenance and using any type hydrocarbons. Sand, which is a very low cost material, is the main component to extract the ceramic to be used in the elements. Although current costs are high, future developments on smaller units will affect the product in a lower price. 100 Kilowatts Bloom energy servers cost around 700.000 USD

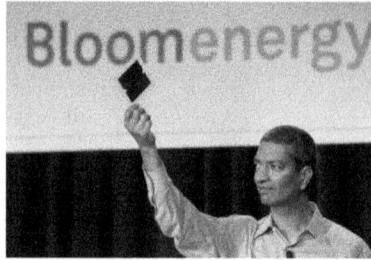

Fig 16-4 KR Sridhar , Bloomenergy's CEO holding a Fuel cell

They claim that their fuel cell, which uses natural gas to make the energy costs really affordable, has a cost of 8-10cents per Kilowatt –hour.

Fig 16-5 Industrial Fuel cell. Source Bloomenergy Inc

Resources:

Fuel cell fundamentals
http://www.fuelcells.org/

Bloomenergy
http://www.bloomenergy.com/

Projects
http://www.lt-automation.com

Growing food in restricted environments

Chapter 17

Lighting

"It is during our darkest moments that we must focus to see the light".

Aristotle Onassis

Vertical Agriculture

The vegetable species require light for their internal photosynthetic process. For traditional (horizontal) agriculture this is achieved by direct exposure to the sun. Implementations of Vertical Agriculture instead, will require either a minimum roof or indoor spaces, which can limit the amount of light a crop will receive. This is why it is important to understand several possible alternatives to supply light to any crop.

17.1 Understanding the Light Spectrum.

The best way to understand how light affects the plants is by understanding the spectrum as shown below:

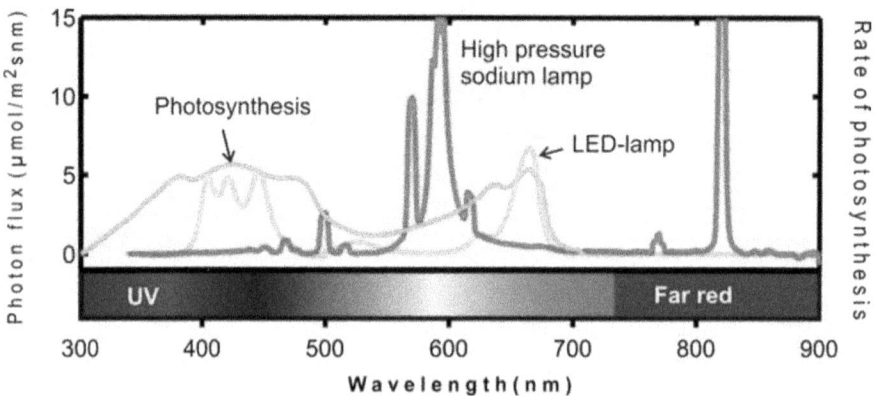

Fig 17-1 Electromagnetic spectrum

The light has components of several colors, each of them with a different wavelength. Some research show that every color and its respective wavelength can have effects on people and plants.

Blue light (420 to 480nm) is equivalent to long sunny days encourages bushy compact growth which makes them perfect for seed starting and leaf growth.

Red (620 to 700 nm) light is similar to the light received during short duration days which help for blooming.

Ultraviolet (UV) (300-420nm) is more important in activating the metabolism and internal chemical production of plants and its further development.

Infrared (IR) (710-900nm) light helps with flowering and growth of plants.

White light is what to human eye sees when there is a combination of the colors red, orange, yellow, green, blue, indigo and violet. White light helps plants to flower.

Chlorophyll absorbs blue and red light . The green light is reflected (or transmitted in other cases), this is why plants look green to our eyes.

In indoor applications, based on the distribution of the plants it will be difficult for some of them to receive enough light to make their photosynthesis. In addition to this problem, the new climate change scenario will bring heavy and continuous rain which can produce long interruptions on the natural light supply so, a sort of artificial light will be required.

When light is controlled near the Infrared light (IR) spectrum, biologists have found that plant cells can grow between 150 and 200 percent faster.

17.2 Lighting

In general there are two possibilities to provide the proper light to the plants:

Natural and Artificial

17.2.1. Natural light

In places with four seasons, sun light can be obtained with restrictions in some specific months. But there are places in the world where sun shines many days of the year so; natural light could be a potential resource to develop agriculture while saving on the energy bill and the usage of artificial light.

Before the 19^{Th} century, daylighting was the only way to provide light to buildings and places. The architecture of the time had good consideration and very creative solutions for supplying light to inner spaces. This knowledge lost its importance when the electric light appeared in the middle of 20th century, until now when costs of energy and lamps, become very expensive for specific applications like vertical agriculture and vertical farming.

It´s not true that all the light, used for vertical agriculture, must be artificial. There are several interesting ways to supply natural light, which allow the possibility to bring sun lighting to inner spaces.

Bringing light to specific locations on buildings will require the use of principles of physics and optics. For transporting the light several methods are used: tubes, pipes, mirror, transparent and hollow devices, etc. This condition has

Growing food in restricted environments

derived in several names like skydomes, skytubes, light pipes, daylight tubes, etc.

Fig 17-2 natural light for interiors. Source : Parans

The quality of light (power, intensity) depends of the chosen technology but there is no much information if it is enough to feed plants and have normal growth rates. Using daylight technology on indoors agriculture application can have a promissory future. The available application starts with simple tubular systems with three main components, as shown below.

The collectors are mostly acrylic diffuse devices which can be either flat or dome type. They allow the capture of sun rays, even in very low angles. In some cases, the collector can rotate to follow the movement of the sun: In this case they are called heliostats. New fluorescent polymeric materials allow light filtering and specific color patterns like green and red, which can be mixed with artificial blue color to generate white light in interior spaces, without the harmful effects of UV light and heat on human health.

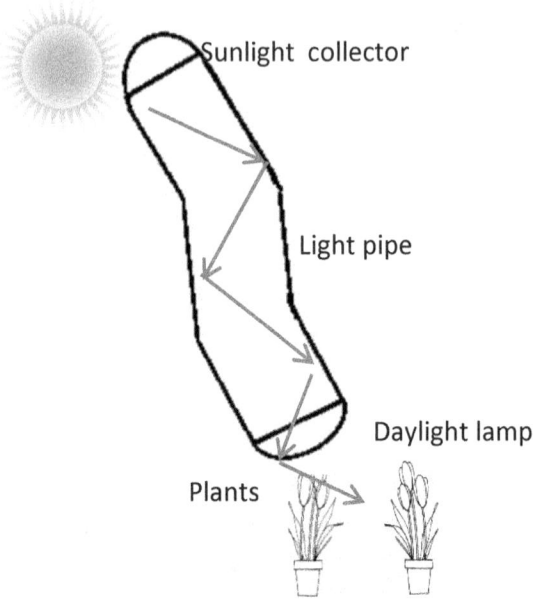

Fig 17-3 Light pipe concept.

The light pipe can be either a rigid or a flexible tube with a laminated and reflective inner surface. With the available technologies the losses of light are around a 40% for rigid tubes. Although flexible tubes are much less efficient, they have a lower cost that makes them very popular.

The outputs are a combination between natural and artificial light. Manufacturers have developed new lamps, some of them with LED technology, which can be used when natural light is not available.

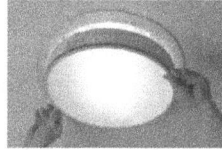

Fig 17-4 Cold Climate Lens Kit Courtesy of US Sunlight Corp Inc

The light tubes can be losing efficiency if they have breaks or they are not completely straight. Humidity can be also problematic because it can change the reflection rates of the inner material. This condition limits the application in some countries where humidity on the air is too high.

In countries with seasons, the heating systems of homes can have losses because the light tubes are path for the heat from the home interior to the cold outdoor environments.

Using the same principle and based on the advancements in the communication technology, more sophisticated systems have been launched to market: Daylight collecting systems with fiber optics technology. The Fiber optics is a small, flexible and few millimeters diameter tube that transmits light and in this case daylight, long distances with very high efficiency and minimum losses.

New manufacturers offer fiber optic solar lighting technology for hybrid applications. The have built solar panels, formed by thousands of fiber optics, which capture the light and transport it to many different places into the buildings' interior, as shown in the picture below.

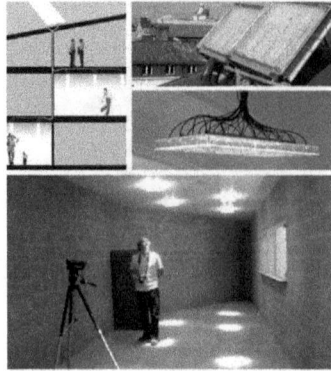

Fig 17-5 Inner places illuminated by Fiber optics panels and natural light.

Numerous applications are available. They include all type of indoors areas where high quality natural lighting is needed, like in indoor agriculture. Some successful projects, where daylight is captured to grow algae in confined spaces, represent a promising technology to save costs on indoor agriculture projects. Extra lamps supply artificial light when illumination is needed and sunlight is unavailable.

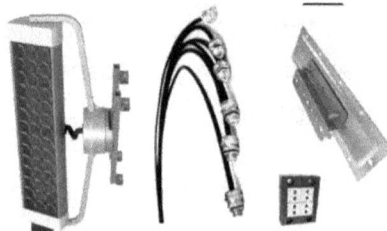

Fig 17-6 Receiver, Fiber Optics and Luminaires.
Parans

17.2.2 Artificial light

Most of the problems for illumination on indoor agriculture applications, except for the associated costs, can be solved by using artificial light obtained from the recently called "grow lights".

Photosynthesis is a process of converting light energy , in the ranges of 400nm to 700nm, into chemical energy and storing it as sugar. Chlorophyll is an abundant pigment of plants that mostly absorbs blue, red and far range light, but not green light.

Some technical parameters are important when selecting the right lamp for your application.

17.3 Parameters to check on Lamps.

- **Photosynthetically Active Radiation PAR**

   PAR, is the spectral range of solar radiation (from 400 to 700 nm) that plants use to develop photosynthesis. Since artificial light is going to be used to replace the sunlight it becomes a very important parameter to select the kind of lamp.

- **Duration**

   The lamps are going to be working most of the time for artificial lighting of indoor agriculture applications. This condition limits their life span so the duration of the lamp becomes an important factor. Some lamps can almost double their duration if working at much lower power than the specified by the manufacturer. This parameter is also related to the power quality available at the installation place, because duration, costs and

Growing food in restricted environments

investments are important when there are potential electrical risks which can affect the lamp operation.

- **Spectrum**

This spectrum refers specifically to the range of the electromagnetic spectrum required for the photosynthetic process in plants. .Some lamps are more suitable for specific needs of color lights. Blue, Red UV and IR are considered important for plant growth.

- **Heat**

Heat emission is a very important parameter because it is related to the energetic efficiency of lamp. From the energy conservation point of view heat is considered as loss and it has effects on the plants because, depending on how distant form the plant they are, direct heat can adversely affect growth.

- **Power**

This is, without doubt, one of the most important selection parameters. The power (the number of watts) reflects the operational costs of the lamps. The number of kilowatts per hour is what everybody has to pay in his power utility bill. The more power you require to perform a job (lighting in this case) the higher the costs you have to pay.

The following table summarizes the technical features for grow lamps.

Lamps	PAR	Duration (Hours)	Spectrum	Heat	Power
Incandescent		1.000		High	High power
High Intensity Discharge		1.000		high	Extreme high power
Fluorescent		20.000	Blue	Low	Low power
High Eff Fluorescent		20.000	Blue, Red	Low	3X lower power
LED		60.000	Blue, Red	Extrem. low	Lowest power

Fig 17-7 Grow lamps comparison chart

17.4 LED Lamps.

Since LED lights are one of the newest products used on growing plants it´s important to understand some aspects about them.

A LED is a solid state device which forms part of a bigger group of semi-conductors named Diodes. They allow current flow in one direction. The word LED comes from Light Emitting Diode, and differentiates from the rest of diodes because of the capability of light emission in different visible and non-visible wavelengths. Recent discoveries allowed the development of blue and red LEDs which work in the specific wavelengths required by plants.

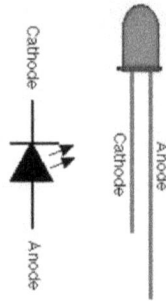

Fig 17-8 LED diode and symbol.

The led lamps are formed by several LEDs connected in combinations of series or parallel branches (see appendix)

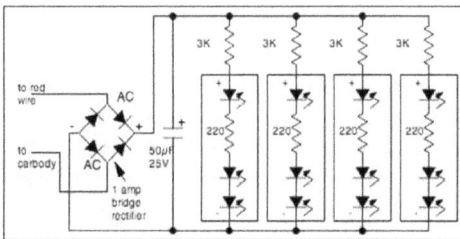

Fig 17-9 Grow LED light electrical diagram

Some research from NASA in his program of Advanced Exploration System (AES), directed by a group of plant biologist, is looking to determine the best ways to grow food in space exploration missions. For the experiments they have used a red leaf lettuce called "outredgeous" and radish plants. Both species were submitted to different light sources: Broad spectrum fluorescent lamps and red and blue LED lighting.

They are still conducting new tests but so far results have shown that LED lights have several advantages in durability, long life and efficiency versus other type of lighting sources.

Growing food in restricted environments

Regarding the plants, several considerable differences were found: The LED lights enhanced the content of anthocyanin, which is a powerful antioxidant which will help space explorers to protect from cosmic radiation. Through LED lights, the exact light wavelengths required by plants can be produced and then help them with the production of chlorophyll, which is achieved by the photosynthesis.

Without a very good testing equipment it´s very difficult to determine the quality of the products so, normal buyers of LED lights are subject to the manufacturers' honesty and they have to assume that their products fulfill all the technical specifications described in the catalogue.

Fig 17-10 Commercial LED lights

Growing vegetables from seedling to flower, will require a full-cycle LED grow light . These lamps are programmable to develop specific light patterns or light experiments. They can be used for germination, cloning, vegetative growth and flowering.

These technologies are much recommended for automation of indoor agriculture applications, because of the low consumption and because some of them can be easily adapted for remote or local control.

Some people consider that artificial lighting is prohibitively expensive to be used in agriculture. In some places of the planet, like the poles, the night last six months and the cold conditions make it difficult to grow food for fresh consumption. Since most of the food has to be imported, the costs associated with indoor agriculture and mainly artificial lighting can be comparatively equal to make the application feasible.

A very important case is the space industry. It needs and has the budget for artificial lighting to be used in food production in confined and restricted spaces and long duration travels. Because of this, most of the knowledge for the new agriculture comes from the research gained in space applications.

Even in countries with sunny climate, the risks associated with climate change make the crops pretty vulnerable to heavy rains, intense heat, floods and other external events. The need for year–round food production will force the agriculture to use indoor environments and because of this, even with sun light available artificial lighting is going to be needed.

Artificial lighting requires energy and the depending of the efficiency, the costs are also high. This energy consumption and carbon emissions, for some critics, represent the reasons for avoiding it. Alternative, renewable and cheap sources of energy to power artificial lighting systems are required.

Growing food in restricted environments

The technological advancements on LED industry will have an effect on cost of grow light. The use of LED lights for all type of application but mainly for indoor agriculture will force to a mass scale production which in time will help in price reduction.

High pressure sodium lamps (HPS) were the popular option several years ago, but since power is now an operational restriction because of the high costs of energy, LED grow lamps are becoming the best option.

A 1000 watts HPS lamp can be replaced by 300 watts LED lamp, which represents savings of 70% in the power bill. There are some other advantages like less dissipation heat, more than 60.000 working hours (versus de 9-12 months of the HPS lamp), no mercury content or any other toxic component, high PAR matching to provide the specific spectrum color for photosynthesis and no harmful radiation.

In order to buy the LED growth lights you have to consider the following aspects:

- Energy.
  You should buy the LED arrays which require less power in watts to work. Remember the energy is going to be the amount of power over the time them the less power the better.

- Spectrum
  This parameter defines the spectrum you require for your plants. You can buy full spectrum or simply specific frequencies as found in red or blue LED lights.

- Heat.
  Look for the LED lights with less heat when they are on. This is considered part of the efficiency and the health of the plant.

- Working hours.
  These lamps are going to work continuously then the more hours they can have the better.

- Warranty.
  Lower costs don´t mean a good choice. Lamps can fail and you want to make sure that your investment is somehow protected.

Resources:

Natural light
https://www.parans.com
https://www.ussunlight.com

LED lights
https://www.energystar.gov/products/lighting_fans/light_bulbs/learn_about_led_bulbs
https://www.ledlight.com/

Grow lights
http://www.planetnatural.com/product-category/growing-indoors/grow-lights/grow-light-systems/

Chapter 18

Solar energy

Vertical Agriculture

"I´d put my money on the sun and solar energy. What a source of power! I hope we don´t have to wait until oil and coal run out before we tackle that.." Thomas A. Edison , 1931

18.1 Solar collectors

In some countries, because of its location, sun shines almost every day becoming a privilege. In some others, depending on the season, some heating will be required. For example, crop production during winter, or under severe weather conditions as a result of climate change, will require systems to provide some heating, even if it is in a greenhouse.

Some devices were designed to take advantage of the heat coming from the sun. One of these products is called Solar Collector.

The main purpose of a solar thermal collector is to transform sunlight or solar radiation into usable heat. A transporting medium, like air or water, is used to bring the collected heat to any specific location, or application.

A solar collector has two main parts: The collector structure and the heat absorber.

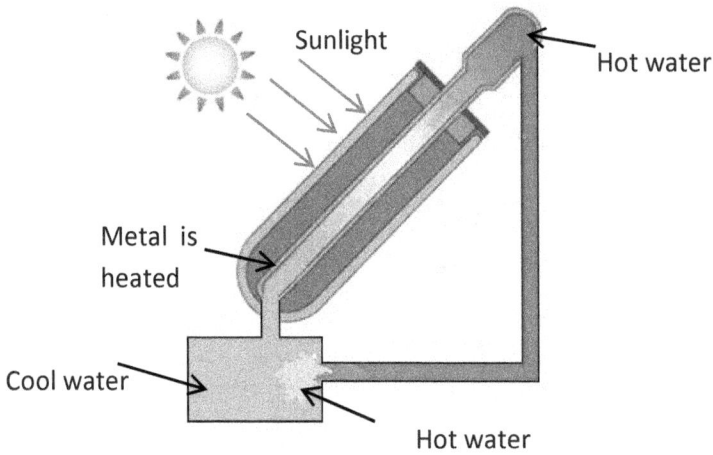

Fig 18-1 Basic working principle of solar collectors.

The surface of solar collectors is usually painted or made of black color, because of its properties of being a color of high degree of absorption of light. Just some of the sun rays are reflected and most of them are absorbed and transmitted to the medium in order to literally "collect" the heat. The relation between the transmitted and the absorbed sun rays is called the efficiency of the solar collectors.

The heat absorber, in modern designs, is formed by a plate in contact with a liquid, inside a two layer vacuum glass tube.

The solar collectors can be classified as:

stationary and sun-tracking.

Stationary models are solar collectors that are located at specific places and with a given and fixed slope. They use a diffuse glass which collects most of the sun rays even in difficult angles. Stationary solar collectors are very popular because they are inexpensive, simple and have easy installation. For most of the applications in agriculture these units are a very good solution.

Sun-tracking solar collectors are far more expensive, because they have an automated control system which helps them to move the surface to the proper angle for maximum heat absorption. The advantage of these collectors is that they are highly efficient, even in low sunlight conditions.

Based on the kind of built-in heat absorber solar collectors can also be classified in: Flat, Tubular or Focus- point. The different types are shown below:

Fig 18-2 Flat Solar Collector

Fig 18-3 Tubular Solar Collector

Fig 18-4 Focus-point solar Collector

The selection of the different types is based on its efficiency and working temperature as shown in the graph below:

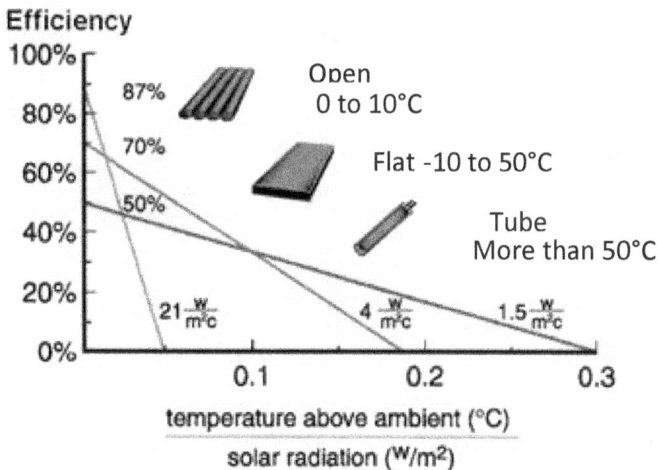

Fig 18-5 Efficiency for solar collectors. Source National Renewable Energy Laboratory, Federal Energy Management Program (FEMP). 2012

A simple application of solar collectors in agriculture is shown below:

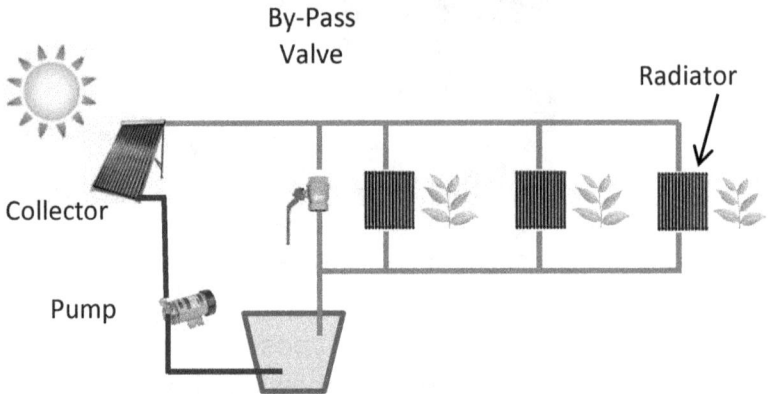

Fig 18-6. Simple application of solar collector systems

A tank with water is connected with a pump to a solar thermal collector. The sunlight heats the water and heats the room. The water is collected again in a reservoir. A by-pass valve can be used if pre-heating of water is required before making transferring heat to the plants.

Before buying any solar collector you should consider the following facts

- Define how much hot water you need. Large tanks are required if you need to cover big areas or you use the water for purposes other than heating.
- Look for the location of the panel. The more are the less sunlight it requires and the better efficiency.
- Choose frost tolerant panels if the cold of your place tends to be extreme.
- Try to avoid electric heaters and pumping systems.

- Evacuated tubes are more efficient than flat panels and require less space. This technology, although very popular, is a little more expensive.
- In some countries rebates and incentives are applicable.
- Look for pricing including installation costs since this could be risky and a little problematic.

18.2 Solar panels

The base of the solar panel is the solar cell. These ones are small cells, which convert sunlight into electricity.

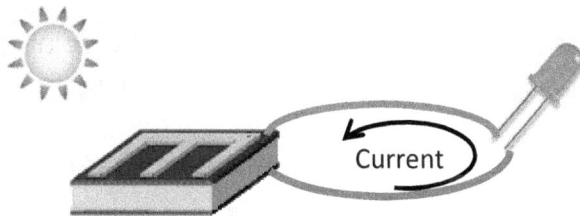

Fig 18-7 Solar Cell working principle

The current is generated by a property called the photovoltaic effect. Depending on the construction material and other technical features, when the sunlight is applied to the surface of the cell there is a flow of electrons. The intensity of the light generates a proportional flow of electrons which is, in time, a current.

Vertical Agriculture

Several solar cells form a Solar Module. A solar panel is an electronic device formed of multiple solar modules, which based on the design, array disposition and construction, can provide a given voltage and current. When several solar panels are involved then the project becomes a Photovoltaic System or simply PV system.

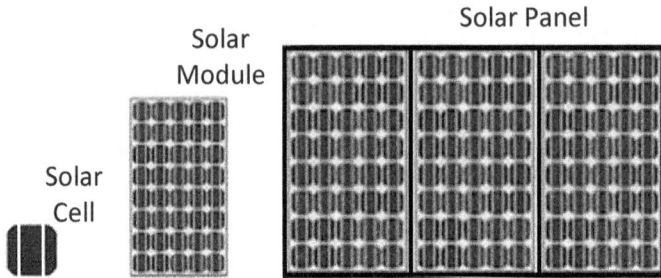

Fig 18-8 From Solar Cell to Solar PV Systems

Since power, according to Ohm´s law is defined as the product of Voltage times Current, the solar panels are specified by its power, expressed in watts.

A solar Photovoltaic system is basically formed by the following :

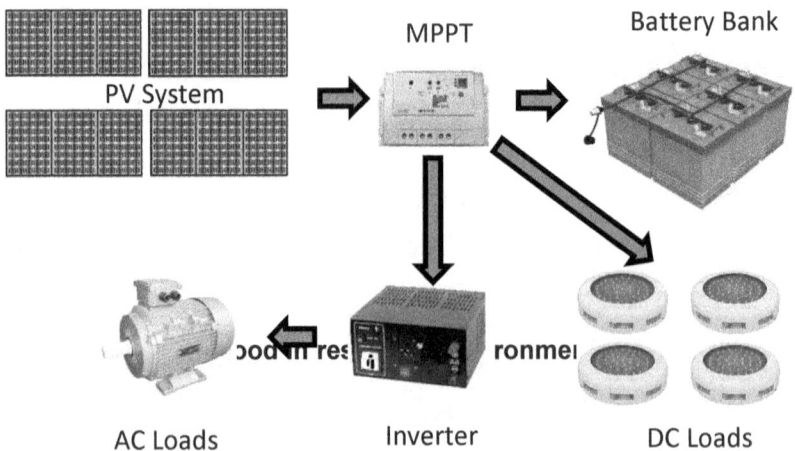

AC Loads   Inverter   DC Loads

Fig 18-9 Components of a PV system

The inverter is only required if the loads are AC type. Some losses of around 10% should be taken into account because of some heat dissipation.

In most cases, it´s possible to make designs without inverters because grow lights, which are widely used for indoor agriculture applications, are considered DC powered devices.

If you want to buy solar panels you have to check the following:

- Make sure that you have space on your roof or any other place to locate the panels. They require certain space and according to the required power the number of panels could exceed the available free space.
- Make sure that you have places with good solar radiation. If the place has shadows it´s quite possible that the efficiency of power generation becomes less than expected.
- Try to use most of the solar energy by running most of the processes during the day.
- Try to use high efficiency equipment avoiding over dimension on the specifications. For example, try to use high efficiency LED lamps instead of incandescent lamps.

- Select a very good inverter with high DC-AC efficiency conversion, if you are going to have AC loads.
- Get an expert to review the technical features of the offered panels and compare them with standards
- Look for certified production (ISO or any applicable local norms). In several cases the operational life is less than expected as the panel can suffer degradation because of use.
- Check for incentives, subsidies and rebates. Solar energy is a help for utilities so, most of the countries encourage the generation of energy with solar panels.

Resources:

Clean energy
http://www.clean-coalition.org
http://www.solarserver.com

Solar Collectors
http://www.solarskies.com
http://www.absolicon.com/

Solar cells
http://www.pveducation.org/

Chapter 19

Control systems and PLCs

"... They have the right to learn. Knowledge must be for all"
Colombian indigenous Karamakate.
From the Movie "Embrace from the Serpent" Ciro Guerra 2015.

Growing food in restricted environments 11

Adapted from the book:
"Fundamentals of Programmable Logic Controllers and Ladder Logic"
ISBN 978-0615800073   by Orlando Charria.

The best way to control production in costs, reliability and operation is through automation. Though any automation process implies a lot of devices, nowadays there are low cost options which represent a lot of future savings because of the advantages it can offer.

19.1 What is a PLC?

The term PLC stands for **P**rogrammable **L**ogic **C**ontroller and can defined as a control device that can be used to provide accurate, reliable and continuous operation of control tasks.

It is used in most of the industrial processes

In automating Greenhouses, Indoors agriculture applications, green building automation, vertical farming and Vertical agriculture projects, a PLC is an electronic equipment that:

- Helps to have a more continuous operation for any agricultural process.
- Drastically reduces the maintenance.
- Simplifies the diagnostic or troubleshooting.
- Helps to protect the life of the operators and the health of plants.

- Guarantees repetitive operational conditions like irrigation or lighting.
- Simplifies the labor for the operator.
- Guarantees an efficient and cost effective process.
- Protects the life of the control elements involved.
- Helps to get a high quality crop production.
- Enhances the production level.
- Raises competitiveness of the companies.
- Aids in the acquisition of current online data.

The electronic design of the PLCs must be made to support the most demanding electrical and mechanical operational conditions such as voltage variations, high temperature, high relative humidity, extreme vibration, etc.

To accomplish such extreme specifications, PLC manufacturers developed technologies that utilize components which, because of their purpose, become more useful. In this race to have a better, more reliable and extreme quality product the result has been a lot of different brands that go to the market with all the required aspects for a technological product:

- Cost
- Service
- Better technical specifications
- Market niche
- Operational environments
- Available options
- Connectivity

A PLC can be used in all type of applications:

- Industrial Machinery
- Home and Building automation
- Robotics
- Automated vehicle guidance
- Conveyor belts
- Plastic injection machines
- Numeric Control for Tool machinery
- Automated storage in Cold rooms
- Air conditioning systems
- Packing machinery
- Redundancy systems in processes
- Water and Energy management systems.
- Industrial Waste and Water treatment plants.
- Substations Energy distribution and transportation systems.
- Parking systems.
- Irrigation systems.
- Car assembly.
- Bottle filling and sealing.
- Amusement parks.
- Automatic lubrication systems.
- Traffic lights.
- Emergency evacuation systems.
- Security systems.
- Vending machines.
- Light and dimmer control.
- Mobile robotics.
- Lighting.
- Administration of doses of nutrient solutions.

For many years, machine manufacturers were used to wire all the control logic using the contacts of relays and their respective coils. This situation led to a very complex and large wiring which was demanding in time and maintenance. Trying to locate a faulty operation was really a difficult task, since the diagnostic tools were scarce.

This wiring was called Relay Ladder Logic (or RLL) mainly because it utilized relays and contacts to provide any control.

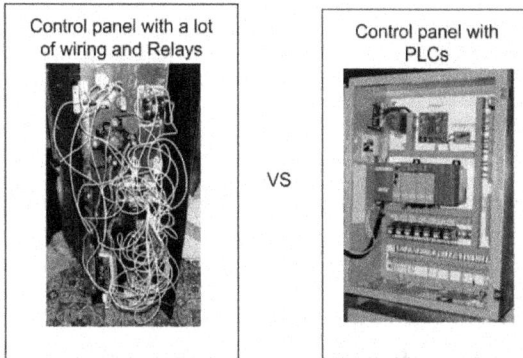

Fig 19-1 Old and new control panels

A very important fact was that there were no many options for troubleshooting. The wire marking technology got extended as one of the most widely used tool to detect a problem. With time, technology on computers brought a new approach.

New processing systems and products appeared to drastically produce changes in control technology that overcame most of the typical problems.

- Many devices involved large and complex control panels.
- Required a lot of work in just wiring the control.
- Difficult to trace a failure in the system.
- To make a change in the control was expensive and difficult.
- Changes had to be made locally, so OEMs and customers had the problem of support.

The microprocessors and its latest descendant, the microcontroller, offered the control designers new possibilities to make faster design and easier to maintain control systems. Very soon, most of the manufacturers incorporated microprocessors in their PLCs, obtaining the following advantages.

- Smaller size and less expensive in all the control equipment.
- Programming software tools for fast development.
- High flexibility to perform changes in control with minimum or no impact in the wiring.
- Powerful diagnostic tools.
- Support of modems, will give you the possibility to change or monitor the control program from any place in the world.
- A more powerful and sophisticated instruction set to perform a lot of difficult calculations and tasks.

The use of PLCs is very popular nowadays. Computers have been trying to get a piece of the industrial automation market, but because of the cost and the industrial characteristic, the final cost is still very high.

Miniature electronic components and the fierce competition of computer technology have turned into modern PCs with more possibility of surviving industrial environments, smaller sizes and really lowering costs. Some PLC manufacturers are adopting this technology to prepare new products that can use the best of both worlds: PLCs and PCs.

19.2 Hardware

For PLC hardware we must understand all the physical components that form a PLC system. Not only the internal electronic architecture. We also want to consider the assembly parts that might be needed to develop automation with PLCs.

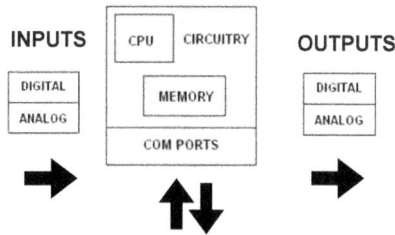

Fig 19-2 Basic PLC´s Structure

The above figure is a basic representation of the PLC parts. The arrows are showing the direction of the information.You can notice that there is no other possibility for bi-directional information than the communication ports.

The PLC cannot change the status of an input or read the information of an element connected to the output. So, INPUTs are used to receive info and OUTPUTs are used to send control commands to the process.

In order to have the better understanding about the way the PLC works, let's take a look to the next graphic:

```
┌──────────────────────┐
│    ┌──────────────────▼──────────────────┐
│    │      READ FROM THE INPUTS           │
│    └──────────────────┬──────────────────┘
│                       ▼
│    ┌─────────────────────────────────────┐
│    │       ANALIZE THE LOGIC             │
│    └──────────────────┬──────────────────┘
│                       ▼
│    ┌─────────────────────────────────────┐
│    │      WRITE TO THE OUTPUTS           │
│    └─────────────────────────────────────┘
└──────────────────────────────────────────┘
```

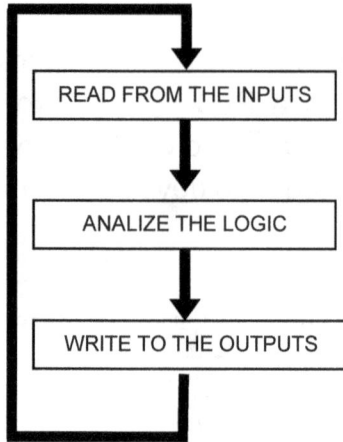

Fig 19-3 PLC´s Internal Logic

The PLC keeps doing this loop consistently. Actually it has been simplified for illustration purposes. The speed of this scanning depends on the CPU itself which is one of the main parameters to determine the real power of a PLC.

If a toggle switch is hooked up, the PLC can only read if it is "On" or "Off" but it cannot force the state to put it, as an example, Off if it's currently ON.

A valve connected to the PLC output can be energized by a PLC command, but if for whatever reason it doesn't receive a voltage needed for the activation, the PLC is unable to determine if the final action is taking place. This final action could be to move a cylinder. A feedback signal could be the solution to this problem. In this case it is necessary to use a device to sense the physical action and then send this signal to the PLC to be read through the inputs.

Growing food in restricted environments   18

As a matter of fact, this is the best way a PLC controls a process: using the sensors and the logic to determine if a proper control action is taking place.

When the PLC only reads the inputs, execute the logic and activates an output then the control system is called an OPEN LOOP.

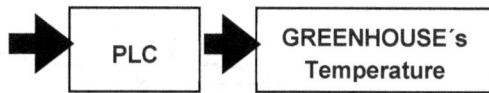

Fig 19-4 Open Loop PLC Control

On the other hand, if the PLC sends the output command and in some way receives a signal that the effects are not the desired ones, then it can make quick corrections until the systems is within the expected conditions. This correction process based on action- effect-result-action is called CLOSED LOOP.

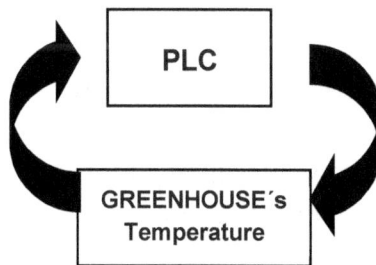

Fig 19-5 Closed Loop PLC Control

In general, every automation project including the applications for indoor agriculture, vertical agriculture (techno) automated greenhouses, hydroponic or aeroponic systems will require a lot of components. The implementations with PLC are a low cost yet powerful alternative. The next graphic shows the different devices which can interact with the PLC.

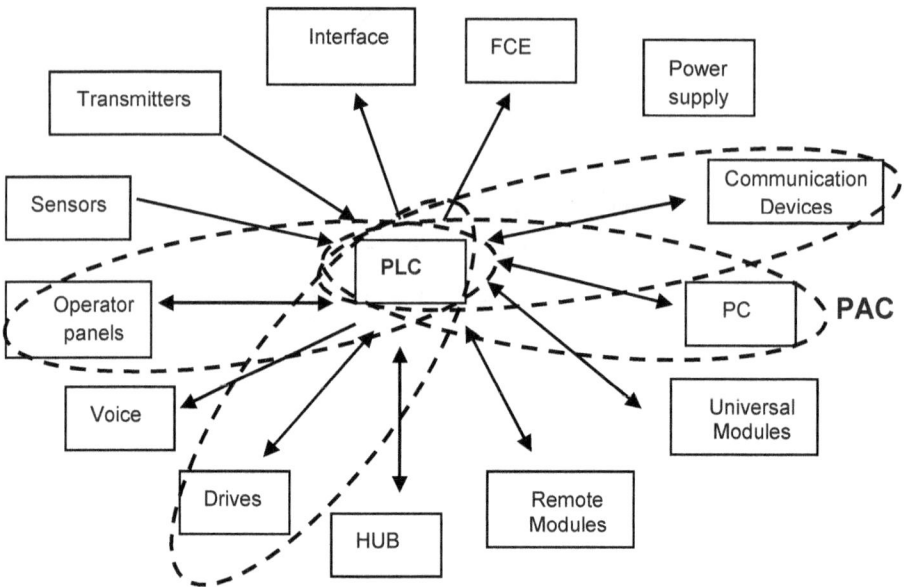

Fig 19-6 Elements for automation with PLCs

A complete book of several pages could be written about all the aspects of every component of an automation project. Every device is a complex industry that involves technology, know-how, marketing, accuracy, etc.

We don't intend to provide a detailed coverage of each automation product, but at least a quick view for you to understand some very common automation situations and the main purpose you could use these devices as part of any Vertical Agriculture project.

To understand the above graphic please take into consideration the following:

- The rectangles represent devices and a complete world of several possibilities. For instance: Drives include all types of commercial drives: AC, DC, Stepper, Servo, etc.
- The arrows represent the direction of the information, hence a two headed arrow means bi-directional information and a single headed arrow means one single direction.
- Dotted lines represent typical commercial combinations. This means that some manufacturers are grouping two different compatible products and creating a new product with more powerful features. This is the competition arena.

Now let's try to have a better understanding of the different components:

19.2.1Power supply.
The power supply is common to all the rest of the elements; this is why it has no connection. Automation devices require a power supply to operate since it provides the specific voltage and current the system requires. If the specifications are different or out of range, more power supplies might be required.

Some power supplies include extra capabilities like: short circuit protection, overvoltage protection, filtering of undesired electrical signals, etc.

You could imagine that there are a lot of types of power supplies, but we will only consider those that are really used in Industrial automation.

Lately, new technologies are allowing the use of an Uninterrupted Power Supply or UPS. These systems can have a battery to provide the power for some given time in case of having an AC power outage.

There are two types of power supplies:

Linear and Switching.

The linear power supplies use a transformer as one of the main devices. The picture below shows the kind of transformer we are talking about.

Fig 19-7 Ferromagnetic transformer

A linear power supply for industrial automation is formed by three basic stages:

Voltage transformation (reduction), Rectifying (changing AC into DC) and Output Regulation (DC voltage in safe levels).

The power supplies that use transformers have simple electrical diagrams as the one below. It includes just few components.

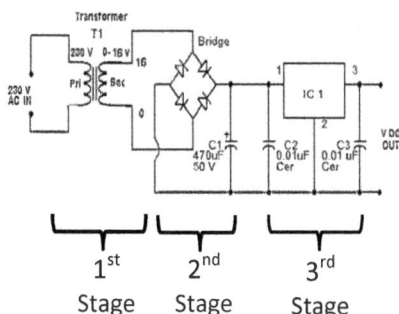

Fig 19-8. Typical linear power supply diagram.

The power supplies that use transformers have simple electrical diagrams as the one above. It has just few components.

The transformers are heavy, bulky and dissipate heat when they are working. This situation puts them in a very non- efficient position when compared to the ones used in switching power supplies.

Today, machine manufacturers prefer to use switching power supplies instead of linear power supplies because of the following advantages:

- Less heat involved so they are more efficient.
- They can absorb any input voltage variation within a very wide range. For instance, they can accept any voltage ranging from 80 to 240VAC.
- Less maintenance required.
- Better output voltage regulation.
- Minimal space required.

- No fuses required since the electronics disconnect the load when short circuits are present.
- Complex circuitry with better operational features.

The figure shows the electric diagram of a very common switching power supply. At first sight you notice that there are more components involved so they are higher technology and more expensive.

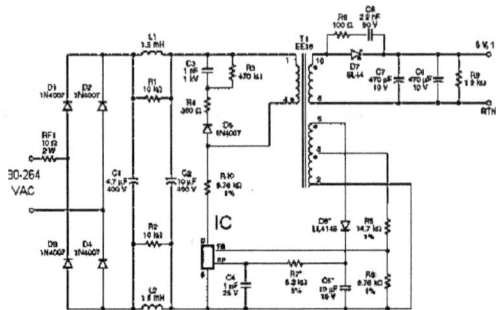

Fig 19-9 Typical switching power supply diagram.

Two components are very important here: The IC, which internally is a very complex device, and the ferrite transformer. This last one is a very small transformer that can work at higher frequencies than the normal transformers.

Fig 19-10 Ferrite

Fig 19-11 Ferrite cores with assorted wiring

The figures in the other page, show a ferrite transformer (left) and assorted types of transformers and coils made on ferrite cores (right).

This is the type of power supply the PLC manufacturers include as an internal option, when offering the PLC with built-in power supply.

Fig 19-12 Switching power supplies.

19.1.3 PLC.

If you observe the main graphic you'll see that the PLC is like the "center" of everything. This was done on purpose, since the PLC can be considered as the core of the automation solution based on PLC. There are some other automation solutions based on PC.

The PLC interconnects most of the devices in several ways. Normally, the information from the automation project is collected, processed, controlled by the PLC. Then the PLC is consulted from other external equipment like PCs, but it's good to clarify that the main purpose of the PLC is only to control the process.

19.1.4 Operator Panel.

The interface between the process or machine (controlled by the PLC) and the person who operates the machine is called Human Machine Interface (**HMI**)

The right selection of the operator panel is very important in order to have an easy, efficient and friendly operation of the machine or process.

You can use from a simple display with text information and two simple buttons up to a complex panel with full color and wide screen which shows graphics of the process in a dynamic way.

Since these devices are very useful to translate the information from the process into understandable and well presented data and graphic trends, most of the automation applications will require Operator panels.

There is a dotted line that groups the PLC and the Operator Panels. This means that nowadays you can get commercial products that include an operator panel and a PLC as single equipment.

Some automation projects will require panels to show just few information from the process. In this case you can use a simple Liquid Crystal Display (LCD).

Fig 19-13 4 Lines x 20 Characters LCD display unit

The displays are referenced by the number of characters per line and the number of lines. For example a display could be 2 lines of sixteen characters each or a 4 lines display with 20 characters for each line, as the one shown above.

There are some cases where you require entering numeric data to inform the machine or process the control settings you want. For example, a set point for the control of a given temperature.

Fig 19-14 Display and Pushbuttons (HMI)

In these cases, in addition to the display information, you need to be able to input the desired control value. For this purpose you can use an operator panel like the one shown below.

When you only want to simplify the number of components involved and want to start or stop a machine, there are some other interfaces available with simpler features, like the one here.

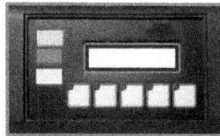

Fig 19-15 Display,Pushbuttons and Indicating lights(HMI)

It only has a display, three indicating lights and five pushbuttons. The program on the PLC will provide the logic to turn on or off the lights and will detect when one of the pushbuttons on the screen has been pressed.

With the new technologies going down in size and scaled up production, you can have more complex operator panels but at affordable pricing. For instance, you can find operator interfaces that can show full color images and not bare text. In some cases you only need to touch the screen to press a button that has been drawn and that is included with the logic of the control program on the PLC.

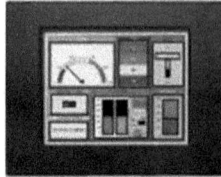

Fig 19-16 Touch screen panel (HMI)

Mini computers and the reduced software environment have launched a new technology in operator interfaces. They are small computers that can run more programs other than just display information. They can really process and handle data bases. The Ethernet ports allow the operator interfaces to be connected to the company network and share information in real time.

Fig 19-17 Touch screen windows CE computer (HMI)

19.1.5 Sensors.

The PLC itself is unable to determine what is happening around. To check for all the variables of the process the PLC is supported by sensing elements or simply devices know as sensors.

The sensors are in charge of the detection of any status change or any new variable value in the environment. As an example, when someone opens a safety door during a process operation a sensor will notify the PLC about this situation. It will have to do the proper actions that can be needed to stop the entire process, sound an alarm, shut off valves, stop motors or simply notify the operator about this normal or risky working condition.

The reliability and accuracy of the sensor are very important selection parameters and must be checked previously, since the whole production can be defective because of a low cost and simple defective sensor.

A final remark about sensors is that the sensors to be connected to PLCs influence the overall operation of the process. It's very important to differentiate from simply a sensor or an industrial type sensor. The term "Industrial", when added to a product, clarifies that the product can work under harsh environments or in some cases it can have multiple operations without a failure. Of course it is understandable that the sensor will fail after some time of operation, but the purpose is to have it operating most of the time.

Below, you will see some examples of digital sensors:

19.1.6 Limit switches

Limit switches are mechanical sensors used to detect when a machine part has reached a certain position or when the machine wants to have a safe operation and the program designer wants to have all doors completely closed.

Fig 19-18 Mechanical Limit Switch

The sensing lever of the limit switch can be selected among a lot of possibilities such as springs, wheels, wires, etc.

19.1.6 Inductive Sensors

19.1.7 In order to avoid mechanical sensors that can be affected by environments full of dirt, water, grease, etc some inductive sensors can be used. The inductive sensor detects metal parts within very short distances (no more than 3 or 4 mm).

Fig 19-19 Electronic Inductive Sensor

For non-metal parts there are capacitive sensors. They are similar to inductive sensors in size and shape.

19.1.8 Barriers

When there is a need to secure the perimeter (like in the case of dangerous machines), some sensor manufacturers offer the so called "Barriers" that are formed by optical (emitter and receiver) sensors. The light beam sent by the emitter can be redirected through mirrors to use only one receiver or simply sensed by several receiver devices. They are the ones that send the detection signal to the PLC.

Fig 19-20 Light Barrier

19.1.9 Reed Relay Switches

Some other applications don´t have harsh environments like the home automation and security systems. They use very low cost devices to detect when windows and doors are closed or open. These are called Reed Relay switches. One of the devices is simply a magnet; the other is a switch that closes when the magnet is exactly in front of it.

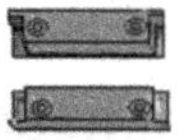

Fig 19-21 Reed Relay Switches

19.1.10 Encoder

To sense the speed of a rotational part, in a digital manner, most of the machine manufacturers use a device named "Encoder". There are just too many different encoder models to mention them here, but in general they can have one output (named Incremental) or two outputs (named Quadrature) to provide certain number of output pulses during 360 degrees of rotation.

Fig 19-22 Mechanical Encoder

19.2 Transmitters.

Several signals, like temperature, humidity, pressure, level, in a automation application are analog, this means that they can have any value. They are different form the other signals are just ON or OFF. When using analog sensors, not all the signals coming out from sensors can be applied directly to the PLC. In some case it is required to change the sensed variable into a so called Standard Signal. A standard signal is a predefined electrical signal that is equivalent to the variable being measured.

The transmitters are in charge of providing accurate information of the variable they measure and supply it to the PLC in an electrical signal free of risks, which can travel long distances without any major distortion of the information.

Let's assume that we have to measure the Pressure of a given system. A Pressure transducer would be the device that we would use to feed the PLC with the info. The Pressure value in any specific unit, is changed by the internal electronic circuitry in the transmitter into an electrical signal (normally a current) which has a specific value within a range, that corresponds to the specific pressure value.

Fig 19-23 Pressure Transducer

When the sensor does not supply the output in a standard signal, then we need to use a transmitter. In this case the sensor is simply a Transducer, changing the variable being measured to an electrical voltage of any magnitude. The transmitter takes this small signal and converts it to a Standard signal. The transmitters are associated with the variable in order to define a name, for example:

K-type thermocouple temperature transmitter or Pt100 RTD temperature transmitter.

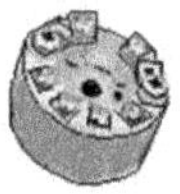

Fig 19-24 K-Thermocouple Transmitter

The transmitters are becoming very sophisticated devices and the sensor manufacturers are adding more and more features. Some recent products can read the variable, store it, show info on a local display, perform some local control if required, and finally send out the signal in the required scaling for a PLC input.

Fig 19-25 Transmitter with Display
and Thermowell Case

19.3   Interface Elements.

The PLC is a control device. This means that its main function is making decisions. For such a purpose there is no need for the PLC to turn into bulky equipment to directly manage high demanding control actions.

For instance, let's assume that the PLC is going to start a really big (in Power) motor so the current it requires, when operating, is above certain limits. Under high voltage or current conditions, the control equipment must be isolated in such a way that the low voltage signal, coming out from the PLC, can drive other elements that are really suitable for these demanding conditions.

19.3.1 Relays

Internally, some PLCs use a small relay which performs the function of isolating the control part and allowing the user to connect loads requiring certain amounts of current.

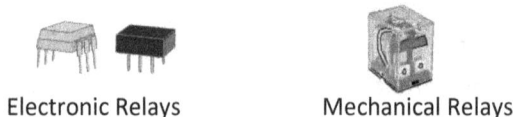

Electronic Relays   Mechanical Relays

Fig 19-26 Relays

Most of the time, these signals are used by the PLC to notify other devices which simply require closing a contact to start an action.

When more current or more contacts are required, then external relays must be used. There are a lot of relay types, with different configuration and purpose, but in general Relays help the PLC output to drive other higher current devices, mostly in cases when the operation could be risky for the electronics in the PLC.

19.3.2 Solid State Relays

The electromechanical relay is a device which has moving parts and because of this, in the time they will fail. This is one of the reasons why the manufacturers of relays have to specify the number of cycles it can make during its life. To overcome this problem there exist the Solid State Relays which use the semiconductor technology and don't have any moving part at all.

Fig 19-27 Solid State Relays

19.3.3 Contactor

A device named **Contactor** is used to drive the high current operation of three phase motors. The control command sent by the PLC is made by sending a low voltage/ low current signal to the contactor's coil.

Fig 19-28 AC Coil Contactor

19.4   Final Control Element.

These elements are devices that receive the control signal from the PLC or the interface elements and execute the desired action in terms of connection, disconnection, aperture, closure, movement, rejection or notification. Their activation reflects itself the final part of the control action.

19.4.1 Motor

The motor is a very common Final control element in charge of producing rotation or displacement of the mechanical part to which it is coupled too. A motor can be driven directly through a contactor or through another device named Variable Frequency Drive. This component will be explained later on, since it forms part of the automation devices.

Fig 19-29 AC Motor

Pumps are similar to motor since they include one.

19.4.2 Valves

Other very popular elements are the valves. These are devices that restrict the flow of any fluid. There are manual, electrical or pneumatic driven valves. In automation we use mostly the ones that can be activated via an electrical signal coming from the PLC.

Since the operational principle is based on electricity they are called Electrovalves. They become Final control elements when they don't require any extra device to perform an action on the process. For instance, water valves, that open or close for water flow, are final control elements.

Fig 19-30 Electrovalve

Electrovalves that send air to another device are not Final Control Elements.

19.4.3 Pistons

The Pneumatic Cylinders or Pistons are very popular Final Control elements. They have two chambers that can be filled with air (pneumatic) or oil (hydraulic) to produce a displacement of the central bar. The valves direct the air or the fluid to one of the chambers.

Fig 19-31 Piston

Assume that we need to control an ejector Piston on a machine. The PLC will send the control signal to a relay, then the relay will energize the coil of an electro-valve and the valve will pass air to the Piston chamber. From this example you can notice that the PLC is an electrical device which is the origin of the signal. On the other hand you see that the Piston is a pneumatic device that requires air to operate and constitute the Final control element.

19.4.5 Actuators

When using industrial valves, there are situations where the torque must be somehow "amplified" in order to perform an opening or closing action. For this purpose you can find the Actuators that are located on top of the valve; based on the direction of the air, they open or close the big valves.

Fig 19-32 Valve with Pneumatic Actuator

19.5   Communication devices

Nowadays, in every automation project the communication is a must. In addition to the communication capabilities of the PLC, some extra communication devices can provide better integration. Within this list we can find:

Modems, Wireless Modems, GSM/GPS/GPRs
Ethernet ports, Serial ports and Web servers

Fig 19-33 Wireless Modem

In some case, some processes are located in distant places where Internet/Ethernet connection is not available. Traditional communication devices like radio or cell phones allow wireless connection to the point where the data can be transmitted through another more popular media.

Some PLC manufacturers have designed optional modules that can be added to the PLCs in case that more communication ports are needed.

We don't intend on going into details about every device since this information is out of scope for this book. Although it is good to know that currently communication is the key to have a very good control system since most of the PLCs will have to interact with a lot of devices ranging from simple devices up to very sophisticated computer networks.

19.6 PC.

The PCs can be considered the universal tool because you need them as a part of your life. In a humble opinion, the normal life in technology can be incomplete if one of two tools is missing: PCs and Internet.

In automation, a PC can be used:

- As a main unit to control a process and this is why it's called PC based control. Some PC manufacturers have developed what they define as Industrial PCs, but their cost, support and software tools have not evolved enough to spread the use in industrial applications.
- As a tool to program the PLC. The PC is used to allocate the software that simulates, compiles, edits and transfers the program to the PLC.
- As the unit that monitors the variables of the process controlled by the PLC. For this purpose   software called SCADA (Supervisory Control and Data acquisition) is used. This software allows a graphical representation of the process to facilitate the

visualization of information and the operation of the elements involved.

- As the tool to troubleshoot the PLC or the process. When something is wrong or there is a need to use specific software to setup or adjust individual elements you can use a PC. This is why it's so important that all the elements used in automation have connectivity with PCs.
- For storing large quantities of process data. The PLC itself has limited data storage capacity which can't be compared to the hard disks on PCs.
- To collect all the information from different processes, PLCs or other devices and then share the information with other computer networks.
- To provide internet access to the PLC. In some cases the PLCs don't have an Ethernet port so they are connected to the PC through serial ports. Once the PC gets communication with the PLC, there are software tools that allow a Remote PC to access the PLC trough this serial connection.

Recently, the manufacturers of PLCs are using small computers to assemble a new device, generating a new technology, which acts as both a PLC and a PC (with some minor limitations). The new name that has been assigned to this device is PAC.

A PAC is a Programmable Automation Controller and some vendors simplify the explanation using the following equation:

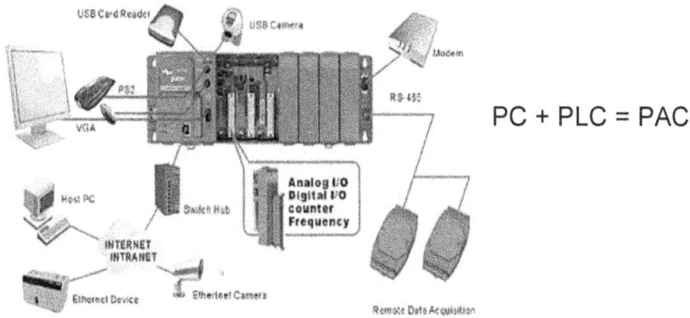

PC + PLC = PAC

Fig 19-34 Programmable Automation Controller (PAC).

From the above graphic you can conclude that a PAC has connection ports for almost any peripheral known for the PLC or the PCs. This feature of having all type of communication and software for data processing is becoming the best reason to use a PAC. This industry is really beginning to grow since its debut to the market five years ago.

The PAC will change the way we perform automation today since its powerful features wide the possibilities of applications, reducing the number of components involved which in time secure the reliability of the whole project.

19.7 Universal modules.

There is a brand new trend to add modules that are multipurpose. The main idea is to make modules capable of reading all type of analog or digital signals.

Fig 19-35 Universal Module

Some manufacturers have launched modules to the market that can be configured to measure either voltage signals (0-1, 0-5 or 0-10V), standard current signals (4-20mA and 0-20mA) or to measure temperature through a Thermocouple sensor in any possible type (S, J, K, T,R, B and more).

The importance of these modules is recognized now. Since most of the time, with older technologies, the PLC users are required to buy different modules for different types of signals and then the result was an increased extra cost of the automation project. Every channel of the module can be set for a different variable reducing overall cost.

The universal modules can be connected as external modules through a pair of communication cables. A lot of modules can be added to the same connection, defining a unique address for each of them.

The cable can have long rungs and can travel all over a place so you can add several modules to this network near the PLC or far away at the end of the cable.

If the modules need to be connected in one of the PLC slots, then we say that the modules have "local connection". These modules limit the size of the PLC but offer an open configuration to the user.

Fig 19-36 Local or Slot Module

Not all the PLC manufacturers offer universal modules and not all of them are interested in producing them since they already have independent modules for each type of signal.

The term "universal" is applied to the programmability of the inputs but nowadays some manufacturers are conceiving that can be applied in a more wide way based on the fact that the communication protocol used (like Modbus) make them capable to be connected to almost any PLC brand in the market. Most of the PLC manufacturers have universal connectivity through this communication protocol.

19.8   Remote Modules.

Supported by their connectivity, the remote modules can act as an expansion for the PLC. The only difference is that these remote modules can be located either on the same panel where the PLC is, hundred meters away or, as an example, in a small town of Paris... provided you have internet connection there.

There are several ways to use remote modules:

- When the PLC is using all the modules and you need to add extra elements, then you add a remote module on a local expansion on the same cabinet.
- When you have a process or greenhouse of considerable distance and you need to run a lot of cabling to the PLC, then you use remote modules to collect all the signals and

then a shielded two wire-cable will be the only requirement to connect the remote modules to the PLC.

- When the company has a LAN (Local Area Network) or wireless LAN and you want to connect the PLC with inputs and outputs signal that can be close to a LAN port, better use a remote module with Ethernet port.

- When a company is interested in monitoring equipment located in distant places (could be in another city or country) a very special remote module could help. The PLC must have an Ethernet port as well.

19.9   Hubs.

When the PLC has an Ethernet port as an option module or as part of the basic architecture, the PLC, for practical purposes, can be considered as a PC.

A HUB is used to provide a communication connection to several devices (mostly computers). This device usually has many of so called TCP/IP ports that allow several devices with TCP/IP port to share a single physical communication channel.

The device avoids that all the equipment connected does not lose communication by assigning them turns, in a really fast speed.

Fig 19-37 HUB

Since the PLC can include an Ethernet port, this is the easiest way to connect the gathered data from the process to a software application anywhere in a Local Area Network (LAN).

The HUB had been a device related more to PCs than to PLCs, but the recent technologies are forcing the manufacturers (including the manufacturers of PLC) to guarantee the best communication available.

19.10 Drives

When an automation application is using any type of motor requiring speed changes, a drive is then required. Since there are AC or DC motors there are also AC or DC drives.

In the case of AC drives it is important to differentiate the Drive from other equipment named Soft Starter. Although application can be similar in some cases, you must know how to make a selection. Pricing is quite the same for small ranges.

When you need to start or stop a motor, and the motor always runs at maximum speed (or Revolutions Per Minute RPM) you should use a Soft Starter. The connection to a PLC is limited to start/stop operation also.

The soft starter connects the motor to the power, producing a ramp on the speed until the motor reaches the maximum allowable speed. If a stop signal is applied then the motor gradually decreases the speed until the motors completely stops.

A Variable Frequency Drive or VFD is a device that allows you to control the speed and the torque of a Tri-phase AC motor in a very efficient way. The big advantage in using a VFD to drive a motor is that the speed can be controlled at any time according to the process needs.

The drive has a special circuitry to protect the motor during operation and to change all the operational parameters such as frequency, times to increase or decrease the speed, operation modes, pre-selected speeds, maximum and minimum speeds, etc. It also has external displays and several feedback signals that can be set to the PLC or other devices for a better overall operation of the process.

The PLC can control the drive through electrical signals that can be: On and Off signals coming from contacts, voltage signals in ranges of 0-10VDC or currents of 4-20mA ranges.

If there is no PLC available the Drive can be operated from a Potentiometer or a small operator panel located on the same equipment.

Fig 19-38 Variable frequency Drive (VFD)

Some manufacturers are adding more powerful capabilities to the drive, and have included a small PLC inside the drive. This enables the PLC to interact with processes and signals in a direct way, simplifying the wiring and lowering total costs.

19.11   Voice Module.

This is one of the latest innovations because you can put your automation process literally to "Speak". You can imagine an automated greenhouse that notifies the operator about problems with pre-recorded words.

Fig 19-39 Voice Module

The voice module is a device with an electronic circuit that can play pre-recorded messages. It also includes a sound amplifier in case that you want to play messages in loud noisy environments like in the case of industrial processes.

The PLC can interact with the Voice module in several ways:

- Using discrete outputs to play few messages.
- Through a binary code using several outputs to play several messages
- Using a serial communication command to inform the module the message to play.
- Stop or play the recording at process' will.

The messages can be recorded using the multimedia software normally found on any computer.

The voice module only requires a power supply and the control signals from the PLC to work. The sound quality is like if you were listening to a CD since the voice is digitized in a very efficient way.

Resources:

Fundamentals of PLC
http://www.amazon.com/Fundamentals-Programmable-Controllers-Ladder-Volume/dp/0615800076

Chapter 20

Electric energy and the Power bill

"Our goal is to fundamentally change the way the world uses energy…we want to change the entire energy infrastructure of the world to zero carbon" Elon Musk, Tesla Motors.

If not now, in future most of the facts point to think that agriculture will be indoors. The crop production will be obtained in closed environments with little or no exposure to the sun light. This new places to grow food will require energy to control lighting, irrigation, temperature, humidity, air quality and support locations like offices will require air conditioning and illumination.

Any project in indoor agriculture, vertical agriculture, vertical farming, automated greenhouse, or controlled environment will require some initial investments. After covering the up-front costs, the electric bill is going to be very important since this will become a recurrent payment which can affect the feasibility and profitability of the whole project. Most of the equipment needed for an indoor agriculture project has been discussed in other chapters. There is good information regarding the criteria to make a good selection for lower operational costs. It's time to explore some ways for supplying with electrical energy to the projects, assuming that you already have selected the equipment with the minimum energy consumption without compromising the efficiency of the system.

In fact, this new agriculture based on artificial lighting for the photosynthetic process in crops would result in food production with higher production costs, since more electric energy is involved, and this last one affects the general carbon footprint. But there is a big mistake in calculations because they are mostly based on energy obtained from power utilities and solar panels.

Vertical Agriculture

One of the main complaints is related to the fact the solar panels require of space, which can be used for food production. In addition some examples mention that the space for solar panels could be up to nearly 9 times the space of the crop, under optimized conditions and high technology.

In addition to the energy supplied by alternate sources like solar panels, there is a big chance that extra energy form the power utility had to be used. This is the reason to consider some important tips that help in minimizing any energy consumption.

Subsidies are popular to encourage a new industry or era like this one, but everybody knows that subsidies won´t last forever and at some point it is necessary to see if the indoor agricultural practice is sustainable in every sense.

If you don´t want to pay a high energy bill you´ll have to look for alternative energy sources or methods because, depending on the size of the crop, lights, air conditioning, pumps for irrigation , filters and monitoring and control equipment will end as a very expensive power bill.

Nowadays, several promising technologies can be found which could help in the sustainability of the solutions and simultaneously allowing lower operation costs. All of them have advantages and disadvantages.

For instance, gas-powered generators allow the generation of electricity which can be used instead of the energy coming from the power utility. Although in most cases and most countries gas is cheaper that electric energy, in additional to the mechanical failure, there are some problems associated with the operation of gas generators: loud noise operation, risks of explosions, and conditioning of places for ventilation.

In some cases, hybrid applications could be the best option. These eco-technologies are applied in other fields, but the more we expand our knowledge on Vertical Agriculture, the more we understand the advantages of using them: Solar (solar collectors, solar panels), tidal, wind, rain, thermal (geo and heat) and flow energy. Another option is the Fuel Cell technology to produce energy.

System analysis.

This analysis must be made locally, at the reader´s location, in order to have a good comparison of costs.

Let´s suppose that 20 Kilowatts are needed to supply power for any application. Some of the options are shown below

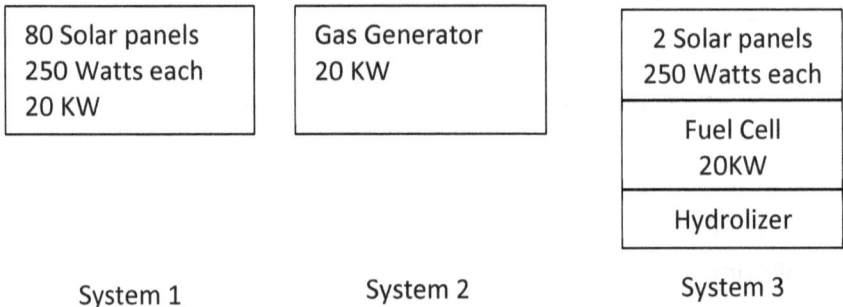

80 Solar panels 250 Watts each 20 KW	Gas Generator 20 KW	2 Solar panels 250 Watts each
		Fuel Cell 20KW
		Hydrolizer

System 1	System 2	System 3

Fig20-1 Different systems for alternative energy generation

System 1. Solar panels.

This solution will require several panels to complete the power requirements. It is assumed that panels produce 24V Dc and not 12, because 12 VDC panels will double the number vof required panels to meet the power requirements.

The same calculations can be made to scale up the power

Some basic calculations are as follows:

20KWatts = 20.000 watts.

If the panel is designed for 250Watts then

Number of panels= 20.000 watts / 250 watts= 80.

The dimensions for each 250 watts panel are

Length:164cm, Width:99,2cm, Height:4,6cm

The area of a panel would be 1,64m x 0.99m= 1, 62 m2

The total area would be = 80 panels x 1,162m2= 129, 6 m2.

In summary, the systems would require a footprint of around 130 m2. This is equivalent to distribute the panels along an area of 10mx 13m.

It´s important to mention that the maximum power output will around 85%.

Some extra components would be an inverter with automatic disconnection of power from the solar panels and connection to the grid. Battery banks could be very expensive too meet the autonomy for very cloudy days and night hours.

System 2. Gas generator.

The gas generators will run with Natural Gas (NG) and Liquid Propane (LP): the efficiency is quite different since the maximum rated continuous power can be affected according to the gas used: LP will provide 20,000 Watts but with NG the systems decreases the output power up to 18,000 Watts.

The Overall Dimensions are:

Length: 121cm, Width:66,5cm, Height:73,7cm

Fuel Consumption at Full Load of NG: is 281 .ft3/hr, while for the LP it will be just 136 ft3/hr .

This equipment works at very high speeds (3600 RPM) so it makes it very noisy. The sound is about 67 dB at 7 meters of distance. Some kind of noise isolation or installation in a distant room is required.

Some extra costs come from a 24V battery, a maintenance kit and cold weather kit if the equipment is going to work in very cold places.

System 3. Hybrid system.

The first important consideration is that although this technology has been developed and applied for several years now, it is not that popular and costs vary from manufacture to manufacturer. The fuel cells are mechanically simple and have no so many moving parts which make them very reliable.

The system is compact and has an overall dimensions :

Length: 199cm, Width:107, Height:60cm

The Fuel used is hydrogen gas 99.99% dry , stored in pressurized bottles. Two bottles will allow generation of 20Kw during two hours continuous, with very low environmental emissions of noise, water and heat.

The two solar panels and the hydrolyzer are used for hydrogen production out of water for an autonomous operation. This also helps in having very low operational costs.

The installations cost arte really low and the only extra equipment could be an inverter.

Recommendations to lower the power bill.

There are some simple recommendations which can help in reducing operational costs when they are considered within the design and initial investments.

- If the construction is already made, some adaptation costs will be necessary. If the construction is about to start, it is suggested to add all the requirements described for the new architecture.
- Buildings must be adapted to have a bioclimatic design without much use of energy. Heating, cooling and ventilation must be as natural or lower cost as possible.
- Try to use as much natural light as possible. Every lamp which can be avoided counts: offices, rooms, corridors, etc. They affect the power bill because of the energy consumption in hours, days, months and years.
- Select components which require less power and have the same work of higher power products. It´s important to mention that sometimes lower costs options can be used instead of very efficient but expensive products. It is a problem of budget and administrative decisions. It is better to pay for some minor extra costs in the power bill rather than buying very expensive products which compromise the feasibility of the project.
- Irrigation must be made with high efficient methods and preferably using gravity rather than pumping. This obligates the designers to consider viable options to manage water and crop distribution.
- Preferably, the water used for the crops must be obtained from rains so collection systems are needed and should be considered as part of the main design.

- Hybrid solutions are much better than simplistic designs with single technological points of view.
- Crops must be part of constant research in order to optimize resources.
- Products must be verified according to their efficiency and manufacturer. The shouldn´t be bought just because the catalogues show the best technical features

Resources:

Projects
www.lt-automation.com

Hybrid fuel cells
http://spectrum.ieee.org/energywise/energy/fossil-fuels/ge-to-muscle-into-fuel-cells-with-hybrid-system

Apendix 1. Types of Agriculture

This is a just a proposal of ways to classify different types of agriculture, which is needed according to the ways it is being implemented.

Readers are encouraged to contact the author to improve this classification.

Based on several criteria, agriculture can be of the any or a conjunction of the following types.

1. **Cosmovision.**
   1.1 Traditional:
   1.2 Modern
2. **Production.**
   2.1 Self-consumption
   2.2 Commercial.
3. **Use**
   3.1 Intensive (mayor desgaste en el suelo).
   3.2 Extensive
4. **Irrigation**
   4.1 Non-irrigated: agua solo es de lluvia y subterránea.
   4.2 Irrigated: El agricultor aporta agua.
5. **Origin**
   5.1 Natural.
   5.2 Transgenic.
6. **Treatment**
   6.1 Ecological.
   6.2 Biological.
   6.3 Organic.
7. **Growing medium**
   7.1 Soilless

7.2 Soil

8. Orientation of Crop
8.1 Horizontal.
8.2 Vertical.

9. Position of the Crop
9.1 Mobile
9.2 Stationary

10. Human Interaction
10.1 Ranchs and Farms
10.2 Plantations

11. Location of the practice
11.1 Urban.
11.2 Rural.

12. Number of Species.
12.1 Single
12.2 Poli

13. Purpose of production
13.1 Edible
13.2 Non edible

14. Control of resources
14.1 Manual
14.2 Technological (Automatic, precision)

15. Applied knowledge
15.1 Ethno
15.2 Techno.

Types of growers or farmers.

1. Big.
2. Small.
3. Hired.

4. Horticulturist
5. Fan.
6. Neophyte.

Fields to categorize an agricultural project:

Location of practice - Orientation- Size- Location-Positioning-Human Interaction-Purpose-Knowledge- Growing Medium-Number of species- Purpose of Production

Vertical Agriculture

Appendix 2

Fundamentals on Electricity

In the World of electric signals two main systems can be considered:

Direct Current (DC) and Alternating Current (AC).

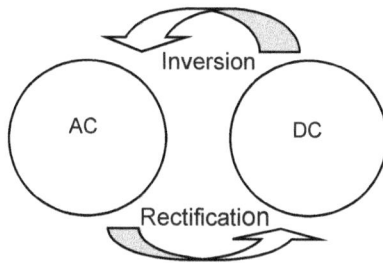

Fig A2-1 Signal conversion.

The process on generating AC from DC is called Inversion and the devices that do this job are called Inverters. On the contrary, to generate DC from AC you use a device called Rectifier and the process is called Rectification.

To have a better understanding of the AC and DC systems, let's assume that we can make a graphic of any electrical signal through time. The DC signal has the same polarity all the time. It can be either negative or positive all the time. When it's changing polarity at any given time it becomes an AC signal. So, for definition: An AC signal is a signal that changes polarity periodically. The speed of these polarity changes is called frequency. You can draw two axis represented by arrows (positive and negative) to show a signal thru time.

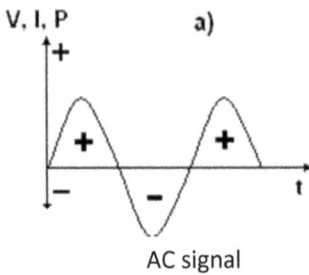

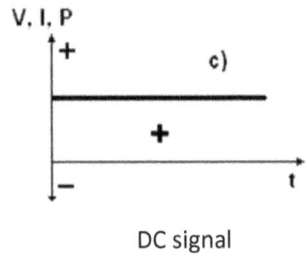

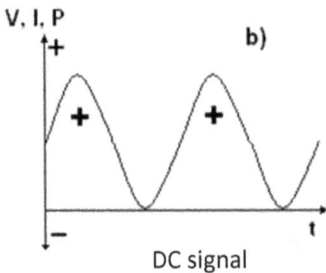

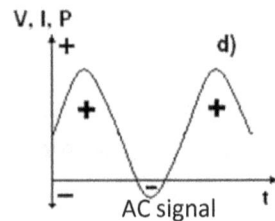

Fig A2-2 AC and DC signals

From the diagrams shown above you can conclude that a) and d) are AC signals and that b) and c) are DC signals. Being under the line of time is a negative signal. Being above the line of time is a Positive signal. Equipment with capabilities of providing energy through AC or DC signals is called Source. Devices that receive the energy from sources are called Loads.

Vertical Agriculture

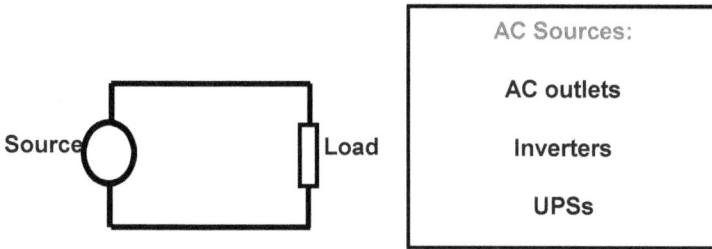

Fig A2-4 Basic electric circuit.

AC Sources:

AC outlets

Inverters

UPSs

Symbols for a Power supply:

Fig A2-5 DC Sources

Fig A2-6 AC Sources

The electrical circuit is a system formed by connecting a Source to a Load, for the energy interchange.

Growing food in restricted environments

3

The Source provides the energy and the Load converts it to work, movement, heat, etc.

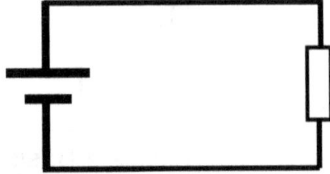

Fig A2-7 DC Source with load

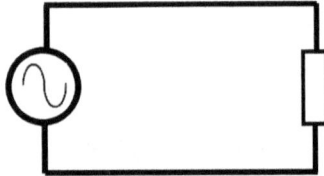

Fig A2-8 AC Source with load

All the electric circuits are governed by the Ohm's Law

"If a Source is connected to a Load, the result is an electron flow called current flowing through the circuit".

This current can be calculated by the following formula:

$$I = V/R$$

I= Current of the circuit in Amps

V= Voltage source in Volts

R= Resistance of load in Ohms

$$I = \frac{V}{R}$$

Ohm´s law

A very good way to learn and remember the Ohm's law is to use the *following triangle:*

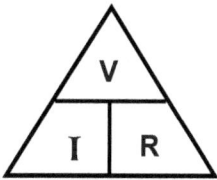

$$I = \frac{V}{R}$$

The idea is that you can calculate any given parameter based on the remaining two.

$$V = I \times R \qquad R = \frac{V}{I}$$

If you need to know the value of any specific parameter, just use your finger to cover it on the triangle. The other two parameters uncovered are a formula that uses the parameters in the exact way they are located: One parameter divided by the other or Current times Resistance if you look specifically for Voltage.

You can see that there is a current only if both terminals of the Source are connected to the corresponding terminals of the Load. In order to control the connection we will introduce a new element named the Switch.

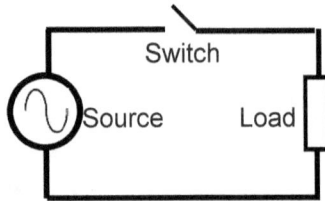

Fig A2-9 Three elements circuit.

We have current flowing through the circuit only when the switch is closed. When this situation occurs you can think that the switch is not present at all.

The three elements (Sources, Loads and Switches) can be connected in the following different ways:

Series connection.

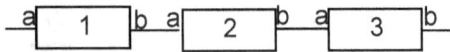

Fig A2-10 Connecting three elements in series.

One end of the element 1 is connected to one end of the element 2. The other end of element 2 is connected to one end of element 3 and so on.

Parallel connection.

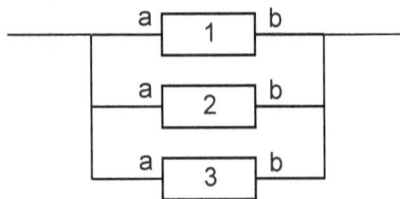

Fig A2-11 Connecting three elements in parallel.

All the "a" terminals are connected together. The same applies to the "b" terminals.

The series and parallel connection can have as many elements as you want. You can make combinations as well.

In some cases you can make connections without paying attention to terminals, but in some others, based on the nature of the element, the resultant electrical circuit is affected by the way you make this connections. In this case these elements are considered as devices with bias or polarity.

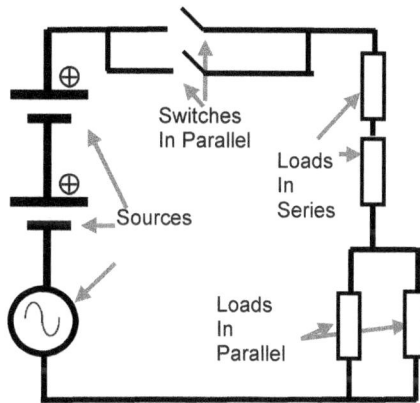

Fig A2-12 A more complex circuit with elements connected in different ways.

In automation, some standards are widely used:

DC Sources:

 12, 24 DC V(DCV stands for DC Volts) with + and – terminals

DC sources are flashlight batteries and car batteries.

AC Sources:

110, 220, 440ACV (ACV stands for AC Volts)

The AC sources use the following names for their lines or terminals

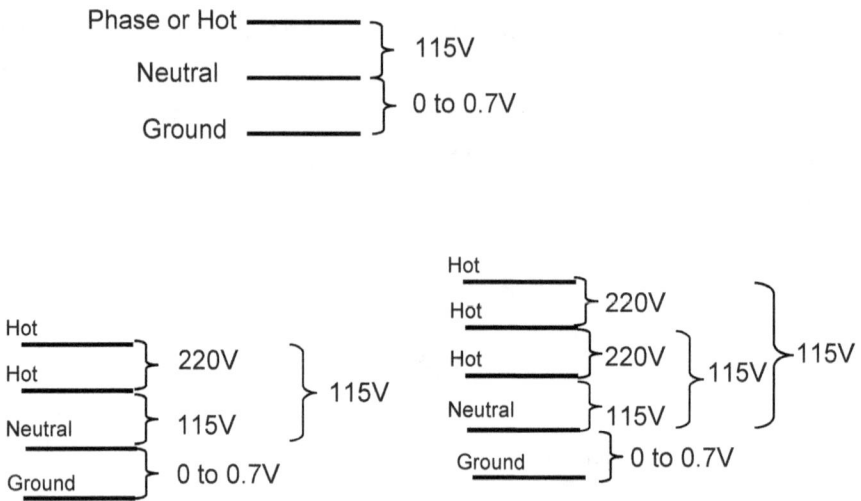

Phase or Hot ———————}
115V
Neutral ———————
0 to 0.7V
Ground ———————

Hot
Hot } 220V
Neutral } 115V } 115V
Ground } 0 to 0.7V

Hot
Hot } 220V
Hot } 220V } 115V } 115V
Neutral } 115V
Ground } 0 to 0.7V

Fig A2-13 Typical configurations of voltages

The Hot wires are also called Phases.

When the load is connected between the Hot and Neutral or between the Hot and Hot you can say that the electric circuit is Monophasic or Single phase.

When a device or load is simultaneously connected to three different Hot wires this is called a Tri-phase circuit.

The Neutral and Ground cables are independent cables in normal installations, but they are connected together before leaving the equipment that acts as the source: The Transformer.

To recognize the terminals on an AC outlet:

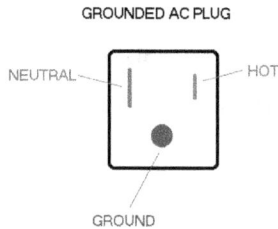

Fig A2-14 AC outlet. Pin names

The standard voltage of any AC outlet is 115ACV in America and 220ACV in Europe.For DC sources, it's assumed that the Red color is + and Black color is -, but always check this since this convention is different in China.

The voltage sources in the ideal condition can supply as much current as required when connected to a load. The truth is that in real conditions the maximum current is limited by several parameters: construction, wire gauges, fuse or breaker protection, etc.

There is a way to relate the Voltage of the source and the maximum current it can provide. This is what we call Power. Ohm's law for power can defined as follows:

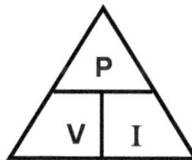

Fig A2-15 Ohms law for power

Power in Watts is the result of multiplying the Voltage times the Current. For a source it's the voltage of the source times the current drained by the Load.

For practical purposes you can use the Triangle to get any parameter based on the other two or simply use the formulas below, where:

I= Current of the circuit in Amps

V= Voltage source in Volts

P= Power of the source in Watts

$$P = V \times I \qquad V = \frac{P}{I} \qquad I = \frac{P}{V}$$

Energy is the power you use in a given time. Normally power utilities will charge you for the number of kilowatts per hour you use (Kw-h or Kwh).

1000 watts = 1Kw

The following energy consumptions are exactly the same 100Kwh:

100 Kw in one hour
200Kw in half an hour
600Kw in 10 min
1200Kw in 5 min

Vertical Agriculture

Power and energy are very important variables because they are associated with operational costs. The less power you can use to have your equipment running the less you have to pay on your power bill.

Index

Vertical Agriculture

Glossary

A

Aeroponics: irrigation technology developed by NASA, which uses mist and bubbles.

Agriculture types: All the possible expressions and categories to develop an agricultural practice

Alternative energy: Energy obtained from different sources other than fossil fuels and other common sources.

Automation: The process of adding control equipment to look for less manual intervention and assure quality and avoid risks.

B

Baily: First person in using the term of Vertical Farming (George Ellis)

Barrier: Protection sensor to secure perimeters and avoid intrusions and risks.

Bioclimatic design: In Architecture, the use of materials for energy efficiency and comfort.

Biomass: Biologic material obtained from living organisms and plants

Biofuel: Fuel obtained from biomass

C

Controlled Environment: Location with controlled parameters like temperature, humidity and pressure.

D

Despommier: A modern spokesman of Vertical Farming (Dickson)

Display: Device to show information from a process and other control functions.

DLR: German aerospace center

Dome

E

Edibles: Material and vegetables than can be eaten.

Energy harvesting: Energy obtained from alternative sources like sun, wind, tides, etc

EPA: Environmental Protection Agency from USA

Ethno: Traditional and ancestral

F

FAO: Food and Agriculture Organization of the United Nations

Food waste: Food that is not consumed and therefore thrown away.

Fracking: Controversial oil extraction technique which uses huge amount s of water

Fuel cell: Device that chemically converts a fuel into electric energy.

G

Gas Generators: Equipment capable of generating electricity by burning a gas.

Greenhouse: Closed and transparent place to isolate plants from external environments.

Grow lamp : Lamp with specific colors and wavelengths for plant growth.

Golden Rules: Proposed set of rules for Vertical Agriculture.

H

Hydrogen: Gas used by Fuel cells to produce electricity

Hydroponics: Irrigation technique that uses a liquid medium.

I

Indoor Agriculture: When all the agricultural activities are made inside a place to protect the crop from the external environment.

Interdisciplinary: Level of group working for better results.

J

K

Kilowatt: Unit of power. 1000 watts

Kilowatt-hour: Unit of energy. Number of kilowatts consumed in one hour.

L

LED: Light emitting diode.

LED Lamps: High efficient lamps made of LEDs

Light Emitting Diode. Solid state electronic device which emits light when a current is applied.

Light Spectrum: Set of colors and wavelengths composing a light beam.

M

Modern Agriculture: Type of Agriculture that can use all available technologies to assure and control crops and food production

N

NASA: Nautical Aerospace Agency from USA.

New Architecture: New constructions made in current days which should address environmental concerns.

Nutritional: Material with positive effects on human health and growth.

Nutrient solution: Liquid medium which is a mixture f water and chemical required for the plants to grow.

O

Operator panel: see Operator interface.

Operator interface: Electronic device that helps operators to perform control function on machinery and processes.

P

PAR : Spectral range of light used by plants to develop photosynthesis.

Phytotron : Plant and tissue growth chamber with controlled conditions of temperature, humidity and light

PLC: Programmable Logic Controller. Electronic equipment which helps to develop control actions with accuracy and reliability.

Power: The consumption in watts of any electrical device.

Propane: A type of gas with application in energy generation.

Pumps: Electromechanical device that helps to bring water to other places and heights with a good pressure.

Q

R

Research, interdisciplinary: When several disciplines are trained and researching on a common topic.

S

Sensor: Electronic device which helps in detecting conditions and measuring variables

Solar Cell: Electronic device which produces electric energy from the sun light.

Solar Collector: Electromechanical device which converts the sunlight into heat.

Solar Panel: Combinations of series and parallel connection circuits of solar cells to provide higher voltages and currents.

Spectrum: Available range of wavelengths of light.

Stationary: That remains in the same position and doesn´t follow the sun.

Sustainable: That can be exploited without depletion.

T

Techno: That allows the use of any technology.

Touch Panel: Operator interface which can react to finger touch with low pressure.

Traditional Agriculture: The agricultural practice performed with minimum tooling and technology.

Transdisciplinary: One of the highest and more difficult levels of group work to guarantee results.

Vertical Agriculture

Transmitter: Electronic device that changes a physical or chemical variable into a standard signal which can be easily connected to PLCs.

U

Urban Agriculture: Agricultural practices carried on cities and the boundaries of rural zones.

V

Vagu: Vertical Agriculture Growth Unit. Low cost system to grow food in Vertical Agriculture.

Vertical Agriculture: Proposed by Charria, this agriculture is directed to grow food under simple rules. It gives equal importance to ethno and techno farmers. Knowledge must be developed and shared by people.

Vertically Integrated Agriculture: A modern and environmental approach to grow food in between transparent walls in buildings

Vertical Farming: A way to prepare and dig the ground to have more space and food (Original idea).

A vision to develop high-tech greenhouses in every floor of buildings to supply food for a growing population (Modern idea)

W

Water Harvesting: Use of water from different sources like rain.

X

Y

Yeng. Modern architect who is famous by his bioclimatic designs (Ken)

Z

Vertical Agriculture

Bioethanol Pilot Plant

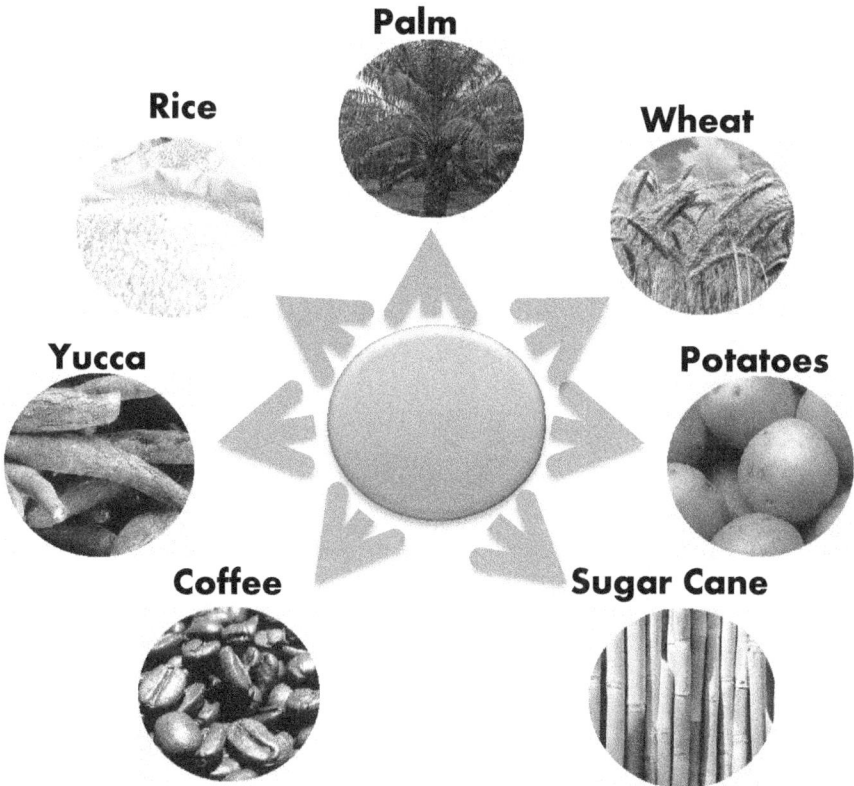

Palm

Rice

Wheat

Yucca

Potatoes

Coffee

Sugar Cane

✓ *Ethanol from biomass*
✓ *Research on other cellulose alternatives.*
✓ *Learn all the processes involved in these technologies.*

Vertical Agriculture

The world currently needs a change to improve the vital conditions. In the quest for oil's substitutes, researchers have been exploring renewables materials like wheat, sugar cane and oil plant, but since these ones are considered food, most of the scientists agree on the fact that food security can be affected. Recently a new trend has increased biofuel from algae (See our biofuel from algae plant).

Our bioethanol production plant is the ideal match for any research on biofuel production from biodegradable material (Biomass) or cellulose in general, allowing you experimentation from different angles.

Research and Education.

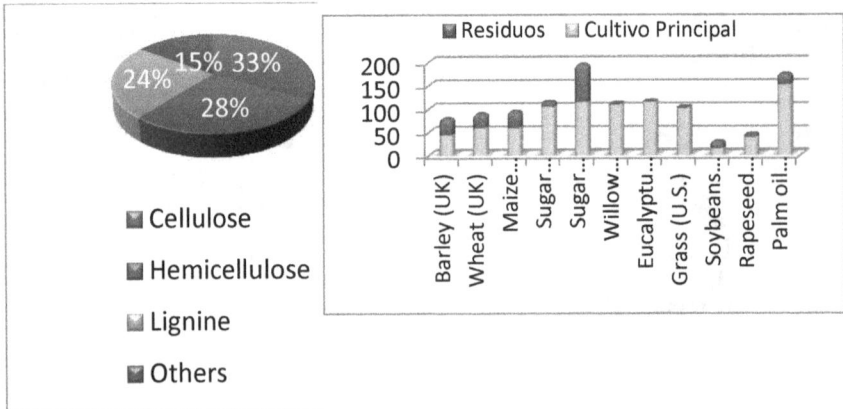

Research

This Bioethanol production plant is able to produce small batches of ethanol from different vegetable raw material, to allow the experimentation and research on efficiency.

Vertical Agriculture

Education

This production plant can be adapted according to the education curriculum train on multivariable control in technical disciplines related to chemical, mechatronic, electrical, electronic or industrial engineering.

One of the big advantages of our educational plant relies on the fact the several groups of students can work simultaneously on the equipment. The control is not limited to a particular brand so you can use multiple controllers or PLC manufactures. However, if you use our PTSF1616[*] you will have a versatile and economic solution to perform both: simple and advanced process control.

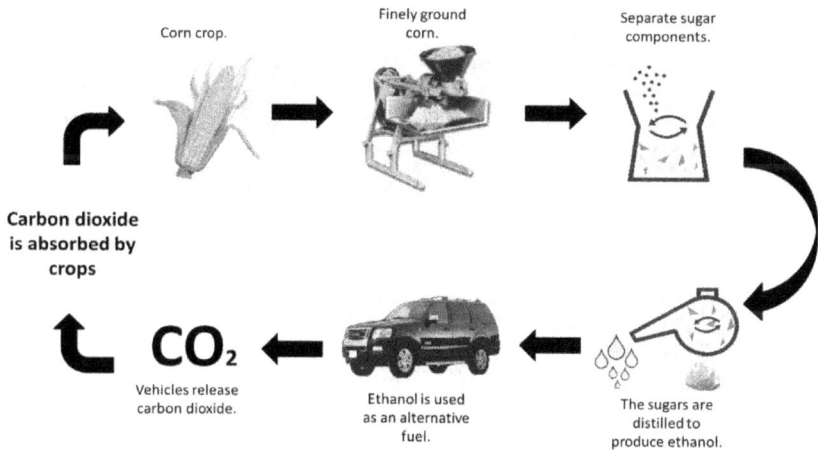

Corn crop.

Finely ground corn.

Separate sugar components.

Carbon dioxide is absorbed by crops

CO_2

Vehicles release carbon dioxide.

Ethanol is used as an alternative fuel.

The sugars are distilled to produce ethanol.

Bioethanol production process

The process of making a fuel depends on several stages, where the control of variables and raw material, are really important to obtain a very good product.

Raw materials for Bioethanol production

The bioethanol can be obtained through processing different vegetables; one of them is the cellulose coming derived from grains, sugar cane African oil palm, and wood, among several more. Another raw material is the starch, which can be obtained from corn, wheat, potatoes, rice, etc. For instance, the sugar obtained from the juice of sugar cane, sweet fruits, can also be used as raw material.

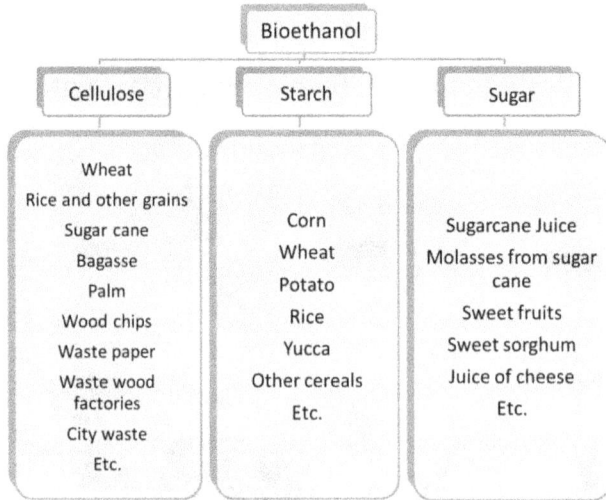

Bioethanol		
Cellulose	Starch	Sugar
Wheat	Corn	Sugarcane Juice
Rice and other grains	Wheat	Molasses from sugar cane
Sugar cane	Potato	Sweet fruits
Bagasse	Rice	Sweet sorghum
Palm	Yucca	Juice of cheese
Wood chips	Other cereals	Etc.
Waste paper	Etc.	
Waste wood factories		
City waste		
Etc.		

Statistics and yields of raw material.

Material	Sugar [%PF]	Culture [T/ha]	Ethanol [L/T]	Yiedl [L/ha]
Sugar Beet	16	60	100	
Jerusalem	16	40	90	6000

Artichoke				
Potato	20	20	120	3600
Grass		2 - 13(MS)	150	2400
Molasses (40 kg / t beet)	50	(2.4)	300	300 - 1950
Corn	58	8.3	390	-720
Wheat	60	5.5	370	3000
Cerum	4.9		23	2040

¿How can you order this production plant?

1. Like a pilot plant.

In this version a control panel handles all the related processes and allows you for any intervention. Through a touch panel you can perform manual or automatic operation, observing all the production variables, diagnose or stopping of the process.

This production plant is made for all of those research ad education disciplines which want to focus on the process directly and not on the control features of every device dudes. It's guaranteed to have an easy, safe and reliable operation.

2. Like a set of control workstations.

The plant is designed no just like an integral process, but like a group of separated sub processes which can be individually operated. The variables, involved in every stage, are channeled toward an individual panel or module where can be connected to our control system PTSF1616[*] or to any other brand controller.

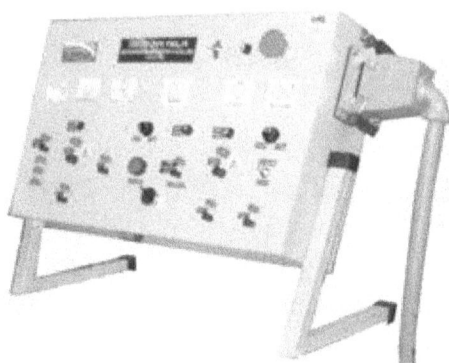

On every workstation, a group of students can develop all kind of control algorithm trough manual, remote or automatic control. Stand-alone or pc control also can be achieved.

This is one of the best educational tools for courses on: process, control, automation, mechatronics, industrial and electronics.

One of the big advantages is that every workstation doesn't affect the others. This is very useful to simultaneously serve a bigger numbers of students. At the end of the course, all the group can collaborate together to run a production batch.

Vertical Agriculture

¿What is included along with the plant?

General	Pilot Plant	Educational Process
• Mash tun • Fermenter • Distiller • Agitators • Activation Valves. • Pumps • Sensors, • Interconnecting Piping. • Water Supply. • Steam supply	• Touch Panel. • Control Panel. • Manual Optional • Data Output Software • Data Acquisition Software • Training courses for researchers.	• PTSF1616[*1] Control Workstation • Connecting cables. • DAS for every workstation. • Up to Four Workstations. • Training Courses for Professors.

What Control Processes Can You Make?

1. Level control and measurement.
2. Stirring and homogenization.
3. Temperature control and measurement.
4. Heating and cooling.
5. Product doses
6. Fermentation.
7. Distillation.
8. Variable control and measurement.
9. PH control and measurement.
10. ON/OFF proportional control.
11. Variables monitoring.
12. Control algorithm.
13. SCADA[*2] Supervisory control (Optional).
14. Caudal rate.

15. Design of safety and intrinsically safe system.

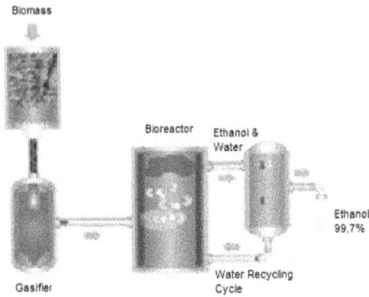

[*¹] See Brochure PTS F1616

[*²] See Brochure SCADA

Latin Tech Inc.
PH 305 848 3517
www.lt-automation.com
sales@latin-tech.net
Miami, USA.

Vertical Agriculture

Climate Chamber-PhytoTron

FIT - 17 - HTL

Really for Climate Change!

Vertical Agriculture

In many research fields it is required an advanced and flexible system to generate Humidity, Temperature and Light profiles.

With this environmental chamber you can:

- Simulate extreme environmental conditions.
- Grow plants and tissues.
- Research on how vegetables species are affected by adverse environments.
- Accelerate tests of climate change on vegetable systems..
- Experiment with artificial light (Red or Blue spectrum) on plant growth.

General Features

- 5" Full Color Touch Screen
- Trends and graphics of variables (T, RH and Light)
- Data logging and downloading to Excel (SD Memory)
- Control from the touch panel
- 15 step profiles for Temperature, Humidity and Light.
- Preset of alarms and notifications.
- Programmable clock and calendar actions.
- Temperature control with gradient: 2 to 47°C ± 1°C.
- Humidity control from 50 to 90%.
- Three temperature sensors located on different positions in the chamber.
- Normal ac outlet connection

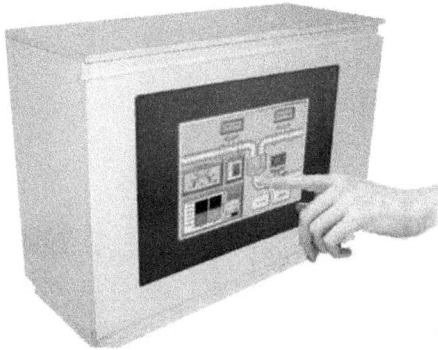

Technical Specifications

Power:	115 V 60Hz
Dimensions:	180 x 71,7 x 78 cm (external)
Weight:	150 kg aprox.
Size:	15 Ft
Capacity:	347 Liters
Cooling system:	Multiflow
Aireation:	Forced flow fan.
Compartiments:	Wire trays, removable.
Isolation:	Rigid Polyurethane foam.
Interior:	Thermo formed Polyestiren.

Vertical Agriculture

Control

Temperature   *2 to 57°C.*
Humidity   *0-99RH.*
Light   *0- Max power.*

Screen

Touch, 5", Color, TFT.
SD memory for easy data extraction.
Clock and calendar
Alarms y events programs.

Optionals

- ○ *Fan for strong wind simulation.*
- ○ *Water Irrigation system for heavy rain simulation.*
- ○ *Higher light or heating levels.*
- ○ *Monitoring Software.*

¿Special requirements?

You can talk to us for special features not included but required .

This is one of the advantages of dealing with designers.

Latin Tech Inc.
PH 305 848 3517
www.lt-automation.com
sales@latin-tech.net
Miami, USA.

Discover and new experience in education and research.
More realistic lab practices.

Wide experience on research, education and industrial equipment.

Other Products

- *Malt, Beer, Soda Pilot Plant*
- *Algae Biofuel Pilot Plant.*
- *Bioethanol Plant.*
- *Inverted Pendulum.*
- *Speed, position and generation plant.*
- *PLC trainer (Generic, AB, Siemens, e.t.c.).*
- *Solar heating system.*
- *Hydrogen cells trainer.*
- *Solar and Wind energy trainer.*
- *Water supply plant.*
- *Motor-generator plant.*
- *Drives.*
- *SCADA.*
- *Other didactic process (in preparation)*
- *Motor-generator.*

Research , Projects or Lectures on

Vertical Agriculture
(English, Español)

Contact us:

verticalagriculture@hotmail.com

Vertical Agriculture

www.ingramcontent.com/pod-product-compliance
Lightning Source LLC
Chambersburg PA
CBHW060329220326
41598CB00023B/2650